Quality of Machined Wood Surfaces

Etele Csanády · Endre Magoss
László Tolvaj

Quality of Machined Wood Surfaces

Springer

Etele Csanády
Department of Wood Engineering
West Hungarian University
Sopron
Hungary

László Tolvaj
Department of Physics and Electrotechnics
West Hungarian University
Sopron
Hungary

Endre Magoss
Department of Wood Engineering
West Hungarian University
Sopron
Hungary

ISBN 978-3-319-22418-3 ISBN 978-3-319-22419-0 (eBook)
DOI 10.1007/978-3-319-22419-0

Library of Congress Control Number: 2015948159

Springer Cham Heidelberg New York Dordrecht London

Printed on acid-free paper

Springer International Publishing AG Switzerland is part of Springer Science+Business Media (www.springer.com)

Foreword

The design and manufacture of wood surfaces are of the same age as mankind. Since ancient times, man has worked out everyday utensils and shelters made of wood. The first time, these objects were dug out from a proper piece of wood, but we cannot speak about furniture in this sense of the word. As the first settlements and towns began to appear the need for furniture continuously increased. The basic elements of furniture (bed, seat, table and chest) were developed by the ancient Egyptians and Greeks. They possessed knowledge about the most important tools and operations to cut boards from a log, the hand planer for making plane and smooth surfaces and the turning lathe to shape cylindrical and profiled contours. In the subsequent thousand years, a relatively slow development took place depending highly on the creativity and wealth of particular nations.

In the second millennium of our time, the development had considerably accelerated, especially in Europe, and different furniture styles were subsequently created and introduced. These "style" furniture were very demanding in material, skill and labour, and, therefore affordable only for a very limited range of consumers.

The past century saw a huge change in the manufacture of furniture. Manual labour was replaced by high performance machines and the design of furniture was also accommodated to the requirements of industrial production. These furniture were already attainable for a wider range of consumers.

In our time we have high performance production systems and, at the same time, a scarcity in high quality raw materials. We are forced to use small logs and wood species not used formerly by the furniture industry. There is great need for new developments enabling to manufacture aesthetic and quality furniture from the available raw material.

The visible surface of any furniture plays a decisive role in its general appearance and judgement. Therefore a proper scientific design and manufacture of wood surfaces became an indisputable necessity in the wood industry. In the world literature there are many contributions to this subject. In order to review, relate, evaluate and assemble all significant facts on the subject, a summarizing book may be of great value to research workers, teachers and application engineers to obtain the existing knowledge in an ordered form. Furthermore, there are interesting experimental results in such a presentation form which are not user-friendly to industrial applications.

This book, *Quality of Machined Wood Surfaces*, is the first attempt to summarize and, in many cases, to complete the existing knowledge on the subject. Several new materials, representation methods and generalizations complete the existing knowledge and making it more usable to application engineers. I would like to emphasize here the importance of only one section: The colour of wood. The colour and gloss are most important aesthetical parameters of wood products, but their practical use in the design and manufacture of wood surfaces is very far from the real possibilities. Effort has been made to change this situation.

I hope this new textbook on the *Quality of Machined Wood Surfaces* will contribute to better understanding of the complicated physical and chemical process associated with the design and manufacture of wood surfaces. The material presented is designed for graduate and postgraduate students, research workers and application engineers working on design and manufacture of wood products.

Sopron, Hungary
June 2015

György Sitkei

Preface

The word *quality*, referred to products or services, has become one of the most frequently used phrases in our everyday life. At the same time, quality is a very complex meaning depending on a myriad of influencing factors pertinent to the given product or service.

In the woodworking industry the parts manufactured for a given end product have almost always free surfaces being in service for many years. These generally visible surfaces should be machined, treated and finished from the properly selected raw material in such a way that they retain their required physical, mechanical and aesthetical properties for a long time. Therefore, problems encountered in the manufacture of wood surfaces deserve special attention.

This book *Quality of Machined Wood Surfaces* is the first attempt to summarize the existing knowledge scattered in the literature and to integrate them into a unified and scientifically well-founded topic. The material published in this book draws on the international literature of several languages, but at the same time it relies on own research works to a considerable extent.

The book consists of an *Overview* and three main chapters. Chapter 2 deals with the general problems of surface stability concerning environmental effects and artificial treatments (irradiation, heat, moisture, mechanical loads). Chapter 3 treats the colour characterization of wood surfaces with a considerable amount of practical applications and new representation methods. The gloss of wood surfaces is a quite *new extended experimental result*, the basic regularities of gloss for natural and treated wood surfaces are established. Chapter 4 of the book deals with the roughness properties of machined wood surfaces using also three-dimensional measurement results. The importance of some, less used distribution parameters is introduced and a quite new topic on the roughness modification effect of wetting is presented.

The authors are especially indebted to Prof. G. Sitkei for reading the completed manuscript and offering many useful suggestions, for instance, in working out generally valid solutions. Special thanks go to Dr. Zoltán Kocsis for offering valuable help in the preparation of illustrations. The authors are also sincerely grateful to the staff of Springer Verlag for their excellent cooperation.

Etele Csanády
Endre Magoss
László Tolvaj

Contents

Chapter 1
Overview

1.1 Introduction

The quality of machined wood surfaces is a primary and important physical and aesthetical characteristic of any wood product. The term "quality" is a very complex and sophisticated notion and the overall quality of a wood surface may be influenced by an array of physical, mechanical and aesthetical properties. Among the physical properties, the surface geometry, the surface roughness, plays a very important role directly influencing a lot of other properties. The same is concerned the photodegradation and stability of surfaces, which also directly affect many important properties. That means that the particular components of the surface quality are in close interrelations with each other and raising difficulties to follow and find out all causal relations encountered in practice.

A further specific feature of the wood surface quality is the relatively high importance of aesthetical properties which can hardly be treated and valued with pure mathematical methods. This circumstance causes some troubles in the process of decision making and product optimisation.

The "Quality of Machined Wood Surfaces" is the first attempt to summarise and analyse the existing theoretical and experimental results scattered in the literature. An Overview and three main chapters give a detailed analysis and description of the components influencing the overall quality of wood surfaces containing many new evaluation and representation techniques, especially in the field of aesthetical properties treated much less in the preceding time.

E. Csanády et al., *Quality of Machined Wood Surfaces*,
DOI 10.1007/978-3-319-22419-0_1

1.2 General Remarks

Looking at furniture of either high-quality or cheaper ones, the appearance of their surface plays a decisive role in our first impressions and appraisal. In particular, bigger plain surfaces are sensitive to any unevenness, even fine scratches which are highly visible in side ways illuminations. Therefore, surface quality of wood products, its evaluation methods, design principles and its optimisation concerning several criteria have with very few exceptions a great importance in wood industry.

The quality of machined wood surfaces is very complex and sophisticated characteristic which should be selected in every case in correspondence with general requirements of the planed product satisfying the utility, aesthetical and cost requirements of interested consumers. That is always a multi-criteria decision process concerning many influencing factors, technological and material availability limitations and, most often, conflicting criteria. Generally, physical, aesthetical and economic criteria should be considered. The physical criteria based on our theoretical and practical knowledge's ensure the best physically possible design. A subsequent assessment of limitation by technology and economics can provide a guide to the development of the appropriate product (surface) with the economically feasible technology and, at the same time, satisfying the requirements of a given consumer range.

From the above discussion, it is clear that in this case we attempt to make the best decision within a specified set of possibilities and the "best" is referred to the simultaneous optimisation of several criteria (Stadler 1988). Therefore, the best means a design within the set of criteria and not the best of all.

Concerning now the machined wood surfaces, their quality may be characterised by the following properties:

a. Surface stability, photodegradation;

   - resistance to UV irradiation,
   - resistance to visible light,
   - resistance to infrared (heat) irradiation and
   - surface layer distortions due to compaction.

b. Moisture relations;

   - contact angle, wettability,
   - infiltration of liquids and
   - moisture exchange with air.

c. Mechanical properties of the surface;

   - surface hardness,
   - resistance to abrasive frictional loadings and
   - surface stability under outdoor conditions.

d. Aesthetical properties;

   - colour and lightness of wood,
   - gloss of wood surfaces,
   - colour inhomogeneity and its pattern,
   - discolouration due to artificial treatments,
   - discolouration due to environmental effects,
   - colour stability under outdoor conditions and
   - polishing ability of wood.

e. Surface geometry;

   - waviness,
   - roughness parameters (R_Q, R_Z, R_k, etc.),
   - standard deviation of roughness parameters and
   - internal defects appearing on the surface.

It should be noted that many surface properties are strictly bound to particular wood species and even to particular wood samples. For example, a round table made of one piece of burl, birds's eye or fiddleback wood sample, is a question of availability. A vivid wood colour or a particular pattern (such as Tasmanian Sassafras) is also restricted to certain part of the world. Beautiful solid wood furniture in natural colours are more common also only in countries in South-east Asia, in the southern part of Western Australia and Tasmania having the precious wood species locally. The European wood species have light colours not much differing from each other. Therefore, the use of wood in its natural colours is restricted to few species (pine furniture).

A machined surface should never be appraised alone on its own; it forms always the free boundary of a product. In certain cases, it serves as a working surface (flooring, kitchen-table) and more often as a free surface with different aesthetical preferences. In outdoor conditions, the outer surface should be resistant to environment effects.

The quality of wood surface will be determined by all important parameters which fulfil the planned utility, aesthetic and economic (costs) requirements and hereby the customer's expectations. An optimum quality decision has now the characteristic feature that in this case there is no other decision for which all the criteria could be improved.

Generally, in the selection procedure of an optimum decision, one can make up a set of criteria of the following types:

- strict preference,
- preference,
- indifference,
- utility requirements and
- constraints.

The main problem of all multi-criteria optimisation procedures is the quantification of the criteria. Concerning surface quality, no multi-criteria analysis is known yet. It is difficult to formulate criteria relations having physical meanings. In particular, aesthetical criteria are difficult to quantify satisfactorily.

Nevertheless, keeping in mind the listed influencing factors and further the preferences, the utility requirements and constraints, the designer can make decision for an optimum or near-optimum surface quality using in a product development.

1.3 Surface Classification

In order to facilitate an optimum decision in selection of proper surface quality, it seems to be useful to classify wood surfaces which may be used in different product developments concerning the demands of consumers.

The following classification is focused first of all on practical uses, several other classifications might have been selected.

S1. Natural surface

It is a debarked surface, according to need, cleaned with high-pressure water jet (young Sugi stems used in interior decorations, table edge made of burl).

S2. Handmade surface

It is used first of all in the earlier time such as dugout items (canoe, trough, pot), carved agricultural and household utensils. The huge wooden column of ancient temples (Φ 1.2 × 20 m) were also made by hand with incredible accuracy. Today, carvings are used for decorative purposes and artefacts. The time consumption for production did not play an important role in the earlier time; in the present-day practice, it is restricted, however, to countries having cheap labour forces (Africa and South-east Asia).

S3. Low-grade machined surface

In this case, colour and roughness have practically no importance; the cost of production is, however, decisive. This requirement is characteristic for back and bottom faces of furnitures as not visible surfaces. Furthermore, railway sleepers, bridges, wharves, boxes, packing cases, post, etc., have similar requirements.

S4. Low demanding surface

Surface colour and roughness are of low importance, production costs on the other hand are of great importance. Examples are sawn or milled surfaces used for platforms, scaffoldings and temporary construction elements.

S5. Utility surfaces

Surfaces in all-day use as flooring, kitchen working tables, writer desk, etc., require a given durability, sometimes antiseptic properties. According to demands, the surface colour and a smooth surface may be sufficient, but not always decisive. An other requirement may be the resistance to chemical reactions or insect attacks. This can fulfil by either a proper selection of wood species (e.g. Syncarpia hillii in sea water construction) or an appropriate surface treatment with transparent wax, oil or lacs.

S6. Demanding surface

Visible surfaces for which the colour and surface roughness have a given importance, the natural endowments of the wood have, however, a limited role. They used for furniture in general use, handles of household utensils (cutlery), etc. Cost requirements play, however, an important role.

S7. Surfaces for high demands

For production of high-quality furniture, utility products and other decorative items and artefacts, but with general availability. Surface finish and colour pattern play an important role; cost requirements are moderate or limited.

S8. Surfaces for extra demands

In this case, the colour and structural pattern of wood material are decisive combined with high-quality finish. The availability and rarity of raw wood material used are also important questions concerning the stability of its value. Cost requirements play a subordinate role (connoisseur pieces).

1.4 Material Classification

Timber properties can be described both qualitatively and quantitatively. Many of the important mechanical properties can give quantitatively which can be used by the designer for proper applications of different wood species taking into account also their growing conditions.

Basically, softwoods and hardwoods are distinguished. The main structural difference between them is the absence of vessels in softwoods. In general, the softwoods have lower density, strength and natural durability compared to hardwoods.

Density is an important characteristic of all timber species indirectly characterises many other mechanical properties indispensable for design purposes. The range of density is depending on the wood species, and it may also vary within a given species as a function of growing conditions. A possible classification according to the density range is given below.

Density range (kg/m^3)	Classification
200–300	Very light
Less than 500	Light
500–750	Moderate
750–1000	Heavy
Over 1000	Very heavy

Within species, considerable density variations may occur. As a rule of thumb, ±(15–20) % difference may be taken into account in relation to the average value.

Very heavy timber species are, for instance, the red ivory (pau-rose)—Berchemia zeyheri; balau—Shorea laevis and S. glauca; keranji—Dialium spp.; Siamese rosewood—Dalbergia cochinchinensis; and Kempas—Koompassia malaccensis.

Softwoods generally belong to the light density range and exceptions are due mainly to extreme growing conditions, for instance, Scots pine growing on the northern territory of Europe.

Hardwoods represent a wide range of density and, therefore, they are generally classified in three groups: heavy, medium and light hardwoods. Heavy hardwoods range in density from about 800 to 1150 kg/m^3 at naturally dried condition. They are naturally durable as they contain within their heartwood extractives preventing the decomposition of the cell wall and which are repellent to wood destroying agents. The sapwood is, however, not durable.

Medium hardwoods are moderately heavy or heavy constructional timbers ranging in density from 650 to 850 kg/m^3. Their durability depends on the conditions where they are used. Many of them, however, are naturally durable in temperate countries where climatic conditions are less hot and humid. Otherwise, they need to be properly treated with coatings or preservatives. Light hardwoods having a density range from 400 to 650 kg/m^3 are the general utility timbers. They are somewhat comparable, more or less, with softwoods in the temperate climate.

The light hardwoods may be excellent for high-class joinery work, cabinet-making, furniture, decorative panelling, flooring, etc. Concerning indoor climatic conditions, they are naturally durable enough for longer life.

In order to save our precious timber species, heavy-duty surfaces, such as parquet flooring, can be designed in multi-layer construction and only the upper layer is made of quality hard timber. For this purpose, the commonly used timber species are the Black locust, Ash, Oak, Canadian hard maple, jatoba, American walnut, doussie, wenge and other Malaysian heavy and medium hardwoods, etc.

The texture of wood materials influences their working and finishing properties. Interlocked grains, coarse and uneven texture due to the presence of abundant wood parenchyma often cause some difficulties in planing, milling, boring and turning, and to produce smooth finish. Planing pinewood in tangential direction, the broader earlywood stripe will often be rougher than the latewood stripe.

It is interesting to note that sometimes the working properties are different for particular woodworking operations. As an example, the wood of Mersawa (Anisoptera spp.) can be planed with somewhat difficulties but it is easy to bore and turn. Some Meranti species, on the other hand, give rough surfaces when turning. Mahogany and teak possess excellent texture and produce very good working properties with smooth finish and luxurious reddish gold sheen.

In certain cases, the original colour of the raw material may be important. Appendix 1 contains a selection of wood species with distinct colour hues. It lists wood species with extremely high density and hardness.

1.5 Surface and Surface Engineering

Generally, a surface is the border area of a machined workpiece to its neighbouring space which may be in our case the environment (air, water) or any other material attached or bonded to the workpiece. Just above this surface, a sudden change in physical or chemical properties exists which has almost always a definite interaction with the machined surface. The interaction takes place in the transition layer at the surface and, therefore, it is customary to distinguish an interface of a body having all important physical, chemical and structural properties.

The term surface quality describes the state and all important attributes of the machined surface which play important roles in the evaluation of the final product in relation to its functional performance and aesthetical value.

The term "quality" is very generic and could apply to any desirable feature of the wood surface. Some particular features of interest may be more specific. For instance, the durability of a surface against damage (wear resistance, photodegradation, resistance to environmental effects) may be referred to more specific term "surface integrity" which is generally used in the metal industry.

A distinct property of wood materials is their colour and gloss. This aesthetical property plays almost always a definite role in the evaluation of the general quality of a given wood product.

Keeping in mind the above statements, the surface is mainly characterised by its topography, colour and glossiness, while interface is characterised by its microstructure (anatomy), physical and mechanical properties and even chemical properties.

Manufacturing a particular product and satisfying a given need, the desired surface will be produced by surface engineering. Latter is almost as old as the mankind. Wood is perhaps the most ancient structural material shaped and used by the mankind. Using hand tools, large wooden columns are worked out with incredible accuracy and excellent surface finish several thousands of years ago.

Also today, indigenous people in many countries make their wood carvings exclusively with hand in beautiful surface finish. Some excellent surface properties of different wood species were recognised several hundred years ago. Let we mention only a few examples: the use of Totara wood for war canoes by the Maori, the same use of Huon pine in Tasmania. The turpentine wood (Fraser Islands, Australia) due to its high silica content is exceptionally resistant to marine borers and has been used for sidings in the construction of Suez Canal and for wharves worldwide. Some high-density species (Guaiacum sanctum, Buxus colchica) having also high-resin content are the most valuable where non-contaminate lubrication is desired. Ship propeller-shaft bearings, food-handling machinery, etc., made such a demand for these wood species that the trees everywhere on the "Endangered List" and mature specimens are very rare.

In many cases, the most important property of a wood surface is its colour and shine. Therefore, there are some very valuable wood species thanks to their exceptional colour hues. Let be mentioned here the Huon pine and satinwood with their beautiful yellow colour, the Tasmanian myrtle with its mid-brown and the Jarrah with its deep-brown colour. Costa Rica's Nazareno (Peltogyne purpurea) has beautiful vivid red colour and the Madagascan "Bois de Rose" (Dalbergia maritima) has an unrivalled deep lilac-red colour.

Today, the surface engineering can rely on the theoretical and experimental results of various science branches such as wood anatomy, wood physics, wood machining and surface science. The surface engineering, as a new discipline, may provide an important mean for engineering product differentiation in terms of quality, performance and cost. The design of the surface for a given set of requirements makes it possible to use cost-effective production technologies and, at the same time, to achieve performance enhancement and good aesthetical values. This activity is highly multidisciplinary, and a successful implementation of surface engineering requires further theoretical and experimental efforts and, first of all, an integrated approach just at the design stage. Manipulating material selection, the sequence of woodworking operations and surface finish, one can achieve optimal surface property for specific applications in the most cost-effective manner.

Surface engineering design may help to achieve desired properties and characteristics of a surface-engineered component including

- general surface topography (waviness, roughness),
- surface stabilisation,
- colour modification in a given range,
- gloss modification,
- improved aesthetic appearance and
- various surface treatments enhancing surface protection, colour and gloss.

A characteristic process in the modern wood technology is the gluing. Coating the surface with veneer is an old technology. Laminated wood plates and glued frameworks, particle and fibre boards are today in common use. Furthermore, the furniture industry widely uses the gluing process for jointing different parts together. The bonding strength of the gluing is highly dependent on the porosity, roughness, waviness and grain direction in relation to the surface. The free surface energy (surface tension) and the acidity of wood play also important role in the gluing process.

Due to the anisotropy of wood, the orientation of the glued pairs in relation to their grain directions has a definite influence on the residual stresses left after gluing and shrinkage stresses due to moisture changes.

The machined surfaces often have various defects which can be attributed to those of the original material and those due to the machining process. The most common material defects are

- cracks with sharp outlines,
- inclusions due to environmental actions such as fire,
- different kinds of knots,
- discoloration due to fungal attack,
- reaction wood and
- irregular growth forms, burl.

An irregular growth (burl) generally reduces the strength of the wood, but the aesthetical value may considerably be enhanced.

The machining process can cause defects in the upper surface layer. The main cause of material degradation is the use of worn tool edges exerting excessive compressive forces onto the surface. In the compressed layer, the cell walls get essentially damaged and, therefore, this layer loses its stability in all aspects. It will have poor mechanical strength and low abrasion resistance; humidity will cause its swelling to various extent giving a wavy surface. Material degradation occurs also under outdoor conditions due to environmental actions such as sun irradiation, moisture and temperature changes.

Material degradations impose a negative effect on responsible parts and structures. The critical level of defects depends on the design requirements and the presence of some defect below a certain threshold may not affect the performance of the product, but its aesthetical value generally decreases (beauty defects).

1.6 Modelling and Optimisation Methods

Manufacturing of any product is an economic activity, and it should be rational in all aspects as much as possible. Design and manufacturing are in close relation and interaction with each other. A purposeful design highly facilitates also the manufacturing of the product.

The manufacturing itself includes several steps such as material selection, process planning, production and integration of the components of products, surface treatment and quality assurance. Further steps are the management and marketing activities.

Concerning wood products, the free surfaces play an important role and they should be designed and manufactured properly and economically satisfying the requirements of the consumers. There are growing needs for quality, process safety, low manufacturing costs and manufacturing times as short as possible. In order to meet the above requirements, optimum design and manufacturing are needed which can generally be accomplished using some kind of modelling and optimisation procedures.

Modelling and optimisation of designing and process parameters on any field of manufacturing is generally a difficult and complicated task. Because of the very many variables and the stochastic nature of many parameters, the development of empirical equations for describing realistic constraint and optimisation criteria, the choice of proper mathematical procedures and optimisation techniques are the main obstacles to perform multi-criteria optimisations. It is no surprise that up to the present no real optimisation procedure for wood surface design and manufacture has appeared in the literature.

Today, more common practice is that the product designer and manufacturing process planner select proper material properties and process parameters using their own experience or those given in publications. Success is highly dependent on the knowledge, ability and talents of designers. To solve these problems, a more effective way would be to develop an explicit mathematical model containing all the functional relationships between input and output and the process parameters determined and expressed analytically.

In order to overcome the existing difficulties, several measures may be applied to find optimum or near-optimum solutions for the design and manufacture of a given product. In paving the road towards an advanced design and manufacturing, the first step may be the work up of a powerful *expert system* containing the most important information on the following subjects:

- material properties for different wood species, availability of materials and their specific costs. Material properties include density, strength, working and polishing properties, colour and colour inhomogeneity, aesthetical value and durability for different conditions.
- Interrelations between design parameters and material properties, and between design parameters and the manufacturing process.
- Principles governing the manufacturing process, the main, mostly empirical relations for the output parameters as a function of operational parameters of machines used in the whole production cycle. The most important output parameters are the accuracy of shape and size, surface quality parameters, including aesthetical values, defined by the designer.
- Economic aspects of the manufacturing process, including the selection of operational parameters and tools for all machining operations ensuring minimum overall costs, shortest production time, high material removal rate and the required quality.

Some progress has been made in the last decades, mainly on the field of the machining process. These refer to one particular machining process such as milling (Bertling 1991) or sanding (Scholz and Ratnasingam 2005) and giving measurement results with useful applications. From the general theory of wood machining, one can learn several useful relationships such as the surface waviness (geometric roughness), surface roughness as a function of cutting speed, tool sharpness and tooth bite for different wood species, cutting force and energy requirement of machining processes, tool wear and tool life (Csanády and Magoss 2013). In searching of an optimum parameter set, it is important that a broad variation of all parameters would be possible. The use of fixed parameters (only one cutting speed, tooth bite, number of edges and edge profile) considerably hinders to find global optimum solution.

Generally, we have several important criteria for optimum design and manufacturing from which one should be selected as the main optimisation criterion. Important criteria are the overall quality, an accentuated characteristic of the product, some aesthetical value, minimum manufacturing costs and short manufacturing time (maximum production rate). The specific energy consumption may also be important but in general other optimisation criteria prevail over the energetic criterion.

The optimisation of a design and manufacturing process is always associated with limitations and constraint in accordance with the product's requirements and machining capabilities. The product's requirements fundamentally influence the following:

- selection of raw materials concerning species, quality and price;
- the necessary machining and surface treatment operations to ensure the expected shape and surface properties. The prescribed surface roughness appears as a constraint in selecting speed, tooth bite and tool sharpness; and
- the demand on the market which determines the scale of serial production.

The existing woodworking machines and equipment supply a range of practical constraints to the given operational process such as

- the available machine tools concerning diameter, width, number of edges and edge profiles;
- machine tool-limiting power, torque, feed force and feed-speed boundaries (upper and lower boundaries of the process parameters); and
- machine limitations in rigidity, capacity, number of operations, accuracy, vibrations and noise.

As outlined above, the design and manufacturing of products is always a non-linear constrained and multi-objective optimisation procedure with an effective optimisation criterion or criteria. Due to the great number of variables and several criteria which can not easily be described numerically, for instance the aesthetical requirements, an accurate treatment of the problem is hardly possible. To analyse similar problems in the metal industry, a large array of optimisation procedures was developed and used (Rao 2011; Fletcher and Powell 1963). The large number of procedures indicates that none of them is generally suitable for arbitrary system solution. Nevertheless, our knowledge in material properties, the main relationships of the machining concerning energy requirements and surface quality, various surface treatments, colour and gloss evaluations, durability tests, etc., allows rational solutions in which the experience and creativity of the designer undoubtedly plays a decisive role. In order to facilitate the work of the designers, it would be useful to process the experimental results in a generally valid form, for instance, in similarity equations (Sitkei 2013).

The design of wood products almost always contains some artistic features which can hardly be designed and optimised with pure mathematical methods. In our opinion, therefore, that the sense of beauty and creativity of designers should always remain the necessary aptitude in beautifying our home and close surroundings.

References

Bertling, R.: Informations- und Expertsystem für die Holzbearbeitung. 8. Holztechnisches Koll. Braunschweig, Workshop, S. 17–31 (1991)
Csanády, E., Magoss, E.: Mechanics of Wood Machining. Springer, Berlin (2013)
Fletcher, R., Powell, M.: A rapidly convergent descent method for minimization. Comput. J. (July 1963)
Rao, R.: Modelling and Optimization of Manufacturing Process. Springer, London (2011)
Scholz, F., Ratnasingam, J.: Optimization of sanding process. In: Proceedings of the 17th IWMS Conference, pp. 422–429 (2005)
Sitkei, G.: Similarity study on the energy requirement of saws. In: Proceedings of the 21st IWMS Conference, pp. 199–205 (2013)
Stadler, W.: Multicriteria Optimization in Engineering and in the Sciences. Plenum Press, New York (1988)

Chapter 2
Wood Surface Stability

2.1 Introduction

Surface stability itself is also a complex property having far-reaching connections with the general resistance to artificial and environmental influences as a function of time. Any deterioration in the physical and mechanical properties of the surface layer, changes in colour and gloss fundamentally decrease the utility and aesthetical value of a product.

The main environmental influences are the light irradiation of different wavelengths, moisture and temperature changes and fungal attacks. Artificial actions such as heat treatment with dry air, humid air or steam also influence the mechanical and aesthetical properties (colour) of the material.

In this chapter, a detailed description and analysis is given concerning the interrelationship between environmental or artificial influences and material response. The most common environmental influence is the UV irradiation causing photodegradation of the wood surface with all its consequences: destruction of physical and mechanical properties of the surface, discolouration, loss of gloss and roughening of the surface. Different artificial heat treatments are also increasingly used to modify some physical and mechanical properties, and also to change colour and colour inhomogeneity. The associated chemical reactions, however, are quite different of those involved in the changes due to light irradiation.

In wood materials, moisture is always present and plays an important role in all environmental and artificial influences. Therefore, the moisture exchange with the air and the moisture movement in wood structures are also treated.

Some mechanical properties of the surface, such as hardness and resistance to friction forces, play also an important role in utility product design. They are also discussed and practical recommendations are given.

E. Csanády et al., *Quality of Machined Wood Surfaces*,
DOI 10.1007/978-3-319-22419-0_2

2.2 General Remarks

Wood is a naturally created polymer composite. The combination of cellulose, hemicelluloses and lignin naturally arranged into tubular structures and eventually forms a cylindrically layered composite; this is the solid wood. Beside these three main components, the extractives play important role in colouring the wood. The unique formations of the woody tissues offer their excellent mechanical and physical properties. Their low density is accompanied by desirable mechanical strength. That is why wood is widely used as structural engineering material for indoor and outdoor applications. In addition, wood is easy to process.

Beside the excellent physical and mechanical properties, the wood has unique colour. Most of wooden colour is naturally inhomogeneous. The colour inhomogeneity of wood is one of the most beautiful colour harmony created by the nature, because its colour range and the colour pattern provide emotive comfort for the observer. The colour hue of wood between red and yellow gives us warm image. The hue angles (see in Sect. 3.2) for wood species are between 45° and 90°. (The 0° hue is red and the 90° belongs to yellow.) There are only a few extremely dark species with hue below 45°. This hue interval is visible by human eye as colour interval between yellow and brown.

Because of its organic nature, wood undergoes abiotic and biotic degradations. The biotic degradation consists of fungal and insect attacks. The abiotic degradations are usually surface phenomena. The abiotic degradation is influenced by the solar radiation, rainfall, heat, humidity content of the surrounding air, presence of oxygen and air pollutants (e.g. acid rain). The colour harmony of wood material is sensitive to light and heat. The most sensitive molecule is the lignin macromolecule because it is good absorber of ultraviolet (UV) light. Chemical analyses showed that the deterioration of light irradiated wood is primarily related to the decomposition of lignin (Tolvaj and Mitsui 2005; Pandey 2005a; Agresti et al. 2013; Tolvaj and Faix 1995; Popescu et al. 2011; Pandey and Vourinen 2008; Calienno et al. 2014). The extractives are also sensitive to light irradiation and heat treatment. Black locust wood changes its colour towards red visibly after only one hour exposure to sunlight. This colour alteration is generated by the degradation of its high extractive content. In contrast, Poplar (having low extractive content) suffers hardly any visible colour change during the same period of time (Persze and Tolvaj 2012).

Wood adsorbs or desorbs water rapidly, depending on the humidity of the surrounding air. The water uptake or loss is accompanied by the swelling or shrinkage of the wood. The rapid surface drying produces surface shrinkage creating cracks. The water uptake influences the colour of wood as well. The increase of moister content produces darker and more saturated colour. In heat or light induced reactions, the moisture content plays also an important role and the moisture always accelerates both photodegradation and thermal discolouration of wood materials.

The density of solid woods depends on species (genetic properties), growing conditions and on age. The density considerably influences several surface properties such as hardness, wear resistance, roughness and gloss properties.

In the following sections, a detailed description of all important questions concerning surface stability will be treated with several new data and practical recommendations.

2.3 Importance and Problems of Surface Stability

The stability as a complex notion expresses any change in the surface properties due to the combined actions of light, heat, moisture and mechanical loadings as a function of elapsed time. These actions may have environmental or artificial origins, and their components and intensities are generally time dependent and of random nature. The loss of stability is associated with serious consequences such as:

- surface material degradation and loss which lower the initial physical, mechanical and aesthetical properties considerably,
- decrease of surface resistance to friction and impact loadings and
- fading and discolouration of the surface.

Wood is a biobased material. Because of this nature, the wooden surface is sensitive to natural decompositions. One of the most important surface parameters is the colour. The colour of the natural woods is perceived by human eyes somewhere between yellow and brown. These colour hues give a feeling of warmth. The machined surface in any anatomical planes represents the colour diversity of the inside lumen surface, the cell wall, the earlywood and the latewood zones. Additionally, the machined surfaces of the different species include lots of micromirrors aligned parallel to the grain. Gloss from these micromirrors gives us more elegant, soft, natural and beautiful texture than that of plastics and metal (Masuda 2001).

The unique colour of wood is sensitive to light, heat and moisture. It is a great challenge for wood scientists to preserve this excellent colour harmony of wood. The colour of non-protected wood changes continuously in outdoor conditions. The surface starts yellowing and the rain leaches out the water soluble chromophoric components. The surface will be more and more grey. In indoor conditions, the colour of surfaces also fades and the dark colours become less saturated. The degradation effect of light can be reduced by transparent UV protective finishes. There is no effective process existing to protect the wood against thermal degradation. It is a slow alteration, but the colour change is visible after a few years.

The surface hardness and wear resistance are also an important parameter. The surface of furniture, claddings and wooden constructions suffer often from different loadings such as friction or impact. The hardness of low-density species is not high enough to protect the surface from damage. There are some transparent surface finishes to increase the resistance to wear or impact loading. The use of these protective finishes is especially important on utility surfaces such as parquet flooring. It is also a common practice that surfaces subjected to high abrasive loadings are made of a thin layer of high-durability wood surfaces.

Wooden products are generally designed for many years in service, during which the main surface properties should be maintained as long as possible. The selection of proper materials, design parameters, the manufacturing process and the surface finish including the necessary protection are the main challenges of any designer.

2.4 Moisture Relations and Wettability

2.4.1 Moisture Exchange of Wood Surface with the Air

Capillaries make wood a porous, hygroscopic material having the ability to adsorb and desorb water. The water content of the wood plays a very important role influencing the physical properties and also the durability of the wood subjected to various treatments and environmental effect.

The most important relations of moisture exchange are the following:

- the moisture content of wood and its variation in time influences, its surface stability, the decay of wood and stability and durability of coatings.
- in internal spaces (living rooms), the wood surface has the ability to equalise the relative humidity of air through moisture exchange with the air.

The wood tends to be in equilibrium with the surrounding environment which generally changes continuously in its temperature and relative humidity. Taking a free water surface, some molecules of water having a given amount of energy can escape from the surface as a vapour into the surrounding air. At the same time, some vapour molecules may re-enter into the water surface (condensation). At equilibrium, the partial vapour pressure above the surface will be equal to the saturated vapour pressure which depends on the temperature.

The saturated vapour pressure can be calculated from the following exponential equation:

$$P_s = 1.3 \times 10^{11} \exp\frac{E}{RT} \quad \text{N/m}^2 \tag{2.1}$$

where

R is the gas constant (8.31 J/mol K),
T is temperature in K and
E $E = 44{,}100$ J/mol is the heat of evaporation

(This value is valid for 20°C and for other temperatures it must be modified.)

The moisture content of wood can be defined and expressed in different ways. The mass of water can be related to the total mass or the dry mass of the wood:

$$U = \frac{m_W}{m_w + m_d} \quad \text{(wet basis)} \tag{2.2}$$

$$X = \frac{m_W}{m_d} \quad \text{(dry basis)} \tag{2.3}$$

where m_W is the mass of water and m_d is the oven-dried mass of the wood. Since wood is a porous material, in certain cases it may be advantageous to use the moisture content by volume:

$$W = \frac{V_w}{V_{ws}} \quad \text{(volume basis)} \tag{2.4}$$

where V_w is the volume of water and V_{ws} is the volume of wood sample. It is clear that the moisture content on volume basis also depends on the porosity n of the material. The different moisture contents are interrelated by the following equations:

$$U = \frac{X}{X+1} \quad \text{and} \quad X = \frac{U}{1-U}$$

and further

$$W = \frac{\rho_{ws}}{\rho_w}(1-n)X$$

where ρ_w is the water density and ρ_{ws} is the volume density of the wood sample.

Due to the cellular structure of wood, the moisture status in wood is a complicated phenomenon. The bound or hygroscopic water is found in the micropores ($r < 50$–100 nm) of wood, first in one layer of molecules and later in several layers. The range of micropores can be approximated using theoretical calculations.

It is well known that at a liquid–gas interface, the surface tension phenomenon exists which is due to intermolecular forces. Over a capillary meniscus with a radius r, due to the water rise in the capillary, the decreased static pressure corresponds to a decreased vapour pressure, which means a smaller relative humidity over the meniscus. According to the Thompson equation, the relative humidity φ over the meniscus is given by:

$$\varphi = \exp\left(\frac{2\gamma\rho_v}{P_v\rho_w}\right) \tag{2.5}$$

where

γ is the surface tension of water (0.072 N/m),
P_v and ρ_v are the vapour pressure and density, respectively,
ρ_w is the water density and
r is the radius of capillary.

Calculations show that $r = 0.36$ nm corresponds to $\varphi = 0.05$ and $r = 100$ nm to $\varphi = 0.98$. In the sorption process, therefore, capillaries over 100 nm radius cannot be filled with water. The amount of water taken up by the sorption process is called the hygroscopic water, and the corresponding moisture content is the fibre saturation point. This moisture content corresponds to saturation of the cell wall without containing any water in the bigger capillaries over radii of 100 nm.

The distribution of microcapillaries in wood is given in Fig. 2.1 showing an interesting picture (Stamm 1946). Most of the capillaries are in the 0.5–1.0 nm range corresponding to $\varphi = 0.2$ and appr. 5 % moisture content. The low relative humidity means a high driving potential (tension) which decreases rapidly with increasing capillary radii. The capillary tension is a real driving force (water potential) according to the following equation:

$$\Psi = \frac{2\gamma\cos\Theta}{r} \quad Pa \tag{2.6}$$

Note that the surface tension slightly decreases with increasing temperature.

Furthermore, the sap in the wood containing soluble nutrients has a smaller surface tension about 0.05 N/m. The surrounding air due to its relative humidity has also a driving potential given by the following equation:

$$\Psi a = \frac{RT}{m_w} ln\varphi \quad bar \tag{2.7}$$

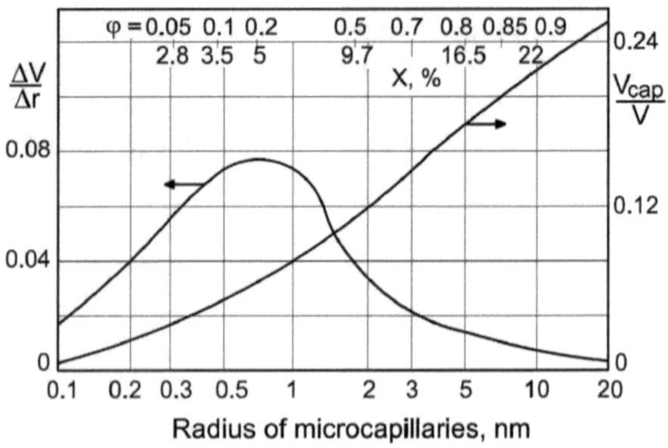

Fig. 2.1 Distribution of microcapillaries in wood

where m_w is the molecular volume of water (18 cm^3/mol) and R is the gas constant (82 bar.cm^3/mol K). An example: the air with 90 % relative humidity at 20 °C has a water potential of appr. −140 bar!

Equilibrium between the wood and the air is only possible if their water potentials are equal to each other. The main advantage of the water potential concept is in that it is valid also for different materials contacting each other and different moisture contents may be in equilibrium, e.g. heart- and sapwoods. The partial vapour pressure difference, as a driving force, can only be used for a homogeneous material having the same physical properties in its entire volume.

The equilibrium moisture content of wood as a function of air relative humidity is shown in Fig. 2.2. The equilibrium moisture content is temperature dependent, and at higher temperatures, it shifts towards lower values.

Due to the shape of the equilibrium moisture content curve, the derivative of the vapour pressure $\Delta P_v/\Delta X$ has a maximum in the $\varphi = 0.3$–0.4 range as depicted in Fig. 2.3, quite similarly to the maximum in Fig. 2.1. Consequently, the shape of the equilibrium moisture curve is dominantly affected by the distribution of micropores in the wood.

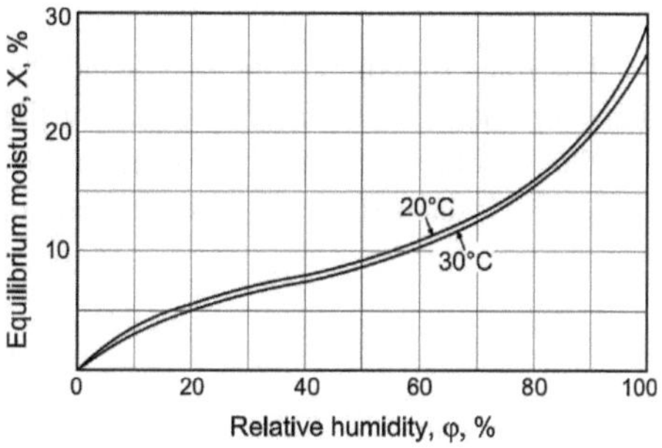

Fig. 2.2 The equilibrium moisture content of wood as a function of air relative humidity

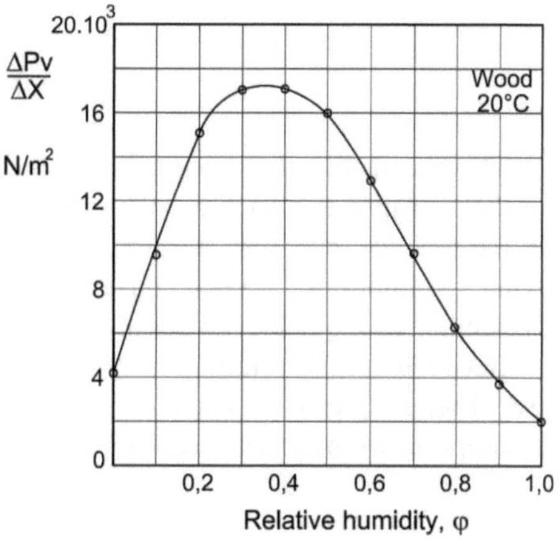

Fig. 2.3 Derivative of the equilibrium moisture content curve

Concerning a homogeneous material, the specific mass transfer on the wood surface is as follows:

$$q = \frac{\beta}{TR_V}(P_s - P_\varphi) \quad N/m^2s \tag{2.8}$$

where

β is the surface mass transfer coefficient, m/s,

P_s and P_φ are the partial pressure on the surface and in the air, respectively, N/m^2 and

R_v is the gas constant of the vapour, m/K (47.1 m/K).

Equation (2.8) may be used in a modified form as follows:

$$q = \beta' \rho_0 (X_s - X_\varphi) \quad kg/m^2s \tag{2.8a}$$

where content corresponding to P_s and P_φ, respectively, and

X_s and X_φ is the equilibrium moisture

ρ_0 is the dry density of wood.

The mass transfer coefficients β and β' have different values, and the interrelation between them is obtained by equating the Eqs. (2.8) and (2.8a).

The moisture transferred to the surface should be conducted into the material which is described by the diffusion equation:

$$q = D\rho_0 \frac{\Delta X}{\Delta x} \quad kg/m^2s \tag{2.9}$$

where D means the diffusion coefficient depending on the moisture content and temperature, m^2/s. The boundary condition is given by equating Eqs. (2.8a) and (2.9) in the following form:

$$\left.\frac{\Delta X}{\Delta x}\right]_{x=0} = \frac{\beta'}{D}(X_s - X_\varphi) \tag{2.10}$$

This means that the process is governed by the β'/D ratio. The accurate solution of the moisture exchange problem is rather complicated and time-consuming task. In the practice, however, a quick estimate is more desirable. Taking some simplifying assumptions, a very convenient solution may be obtained satisfying the possibility of a quick estimation.

Between certain limiting values of the moisture content, the sorption isotherm (see Fig. 2.2) can be taken as approximately a linear function and with this assumption we may write as follows:

$$\frac{dX}{dx} = -k(X - X_e)$$

where X means the instantaneous average moisture content; Xe is the equilibrium moisture content corresponding to a given relative humidity of the air; and k characterises the evaporation rate. Integration of the above equation yields the relation as follows:

$$MR = \frac{X - X_e}{X_1 - X_e} = e^{-kt} \tag{2.11}$$

where X_1 is the initial moisture content in the wood. The left-hand side of the equation is termed the moisture ratio, which characterises the moisture exchange as a decimal number until the equilibrium moisture content has been attained.

The half-time $t_{0.5}$ is termed as the time required for removal of one-half of the evaporable quantity of water, and its value is calculated as follows:

$$t_{0.5} = \frac{\ln 2}{k} = \frac{0.693}{k} \tag{2.12}$$

If the half-time $t_{0.5}$ would be known for different wood species and environmental conditions (temperature, initial equilibrium moisture content and the change in relative humidity), then the moisture uptake or moisture loss can be estimated conveniently using Eqs. (2.11) and (2.12). Furthermore, from the general theory of drying, it is well known that in parallel sided bodies (the infinite plate) the moisture profile follows a parabolic distribution.

With the parabolic assumption, the expected moisture profile for the elapsed half-time $t_{0.5}$ can easily be constructed. Figure 2.4 shows the moisture profile of a 1-inch (25-mm)-thick slab exposed on both sides, for different elapsed times. In the centre line, the decrease of moisture occurs with considerable delay. Due to the parabolic distribution, however, the moisture ratio at the centre line for the half-time $t_{0.5}$ can easily be estimated as 75 %.

The processing of a vast amount of moisture distribution profiles for quite different conditions has shown that the centre line moisture ratio is a uniquely defined function of the dimensionless time $t/t_{0.5}$ as plotted in Fig. 2.5.

The centre line moisture content tends to decrease after an elapsed time between 40 and 50 % of the half-time. Using a half-logarithmic scale, the correlation is not exactly a straight line (exponential function), but the deviation from the exponential law is not essential. Knowing the half-time for a given adsorption or desorption process, the moisture profile distribution for arbitrary time can easily be estimated. It should be noted that the above method is valid only for a uniform initial moisture distribution. With short time variations in the relative humidity, the equalizing process cannot be completed, and therefore, the initial condition for the subsequent relative humidity variations will be rather a parabolic moisture distribution. The latter

Fig. 2.4 Moisture distribution in 1-inch-thick slab

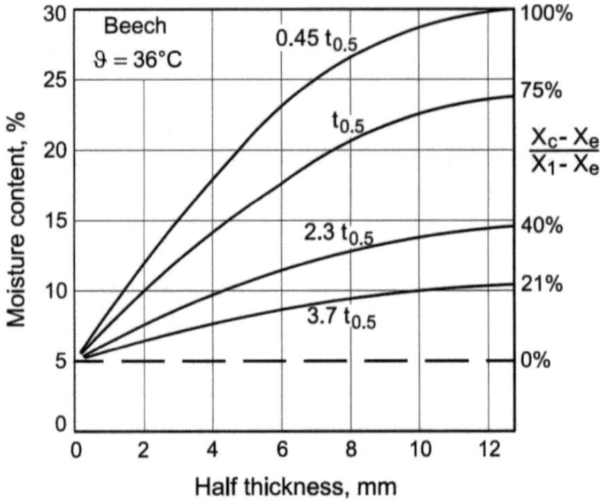

Fig. 2.5 Dimensionless plot for determining the centre line moisture content as a function of time

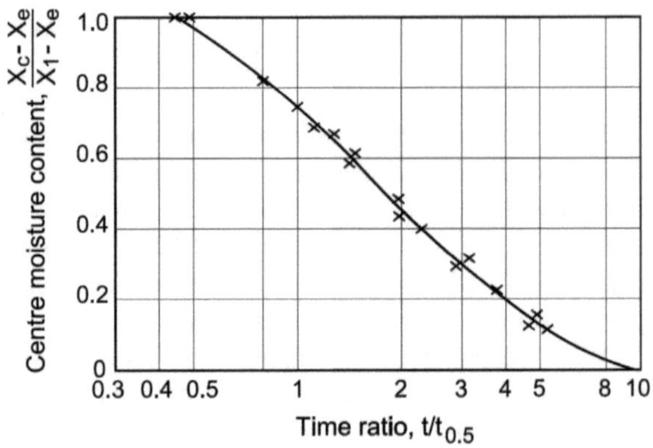

initial condition can be handled theoretically (Лыков and Михайлов 1959), but the solution is quite time consuming and a practice friendly elaboration would be needed.

As Eq. (2.8) suggest, the mass transfer on the surface depends on the partial pressure difference only, but not on the initial value. Equation (2.9) essentially suggests the same. At the same time, carefully conducted experimental results have shown that the adsorption and desorption process can show very anomalous behaviour at high initial moisture contents (Christensen and Kelsey 1959). The transient moisture sorption in wood at high relative humidity is governed also by non-Fickian processes which are associated with a slow sorption in the cell walls. It is also believed that this slow sorption of moisture into the cell wall is due to swelling pressure and its relaxation depends on the creation of sorption sites by molecular bond breaking and formation of micropores. Therefore, the sorption process is governed by two simultaneous different phenomena: the classical diffusion in the cavities and a moisture transfer into the cell walls.

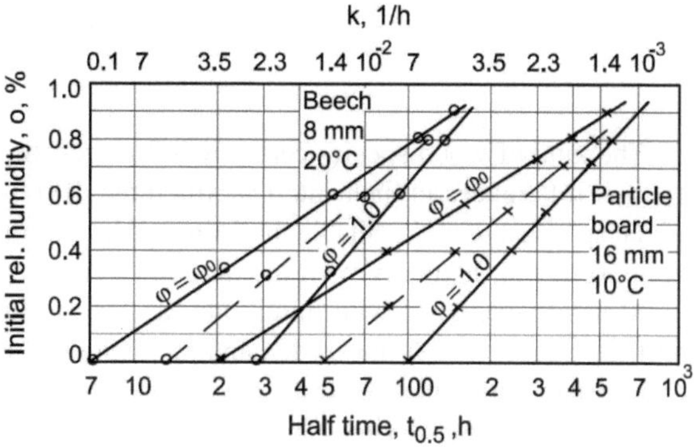

Fig. 2.6 Oblique triangle diagram representing interrelation between half-time and initial relative humidity for Beech and particle board

Instead of seeking a high theory for the above anomalous sorption process, it seems to be more reliable to generalise the existing experimental results. In this case, we will use the half-time with Eqs. (2.11) and (2.12) and Fig. 2.5. Using the detailed measurement results with particle board and several wood species (Béldi and Szabó 1979; Béldi and Bálint 1984, 1986), an "oblique triangle" diagram was found and elaborated containing all measurement results for a given wood species, sample thickness and temperature. Figure 2.6 shows the plot of experimental results on semi-logarithmic scale giving straight lines which refers to the exponential nature of the process. The left line refers to an unsteady step change in relative humidity from φ_0 to φ_1 at which φ_1 only slightly differs from the initial value φ_0. The right line ($\varphi_1 = 1.0$) means the use of saturated air at the given temperature. The broken line gives approximate middle values for $\varphi_1 = (1 + \varphi_0)/2$.

It is interesting to note that the diagrams for Beech and particle board do not differ much from each other if the difference in thicknesses is taken into account.

The effect of thickness is theoretically described by a quadratic function. In practice, however, smaller exponents were systematically found, generally in the range of 1.5 and 1.6. Therefore, half-times for different thicknesses may be corrected in the following manner:

$$\frac{(t_{0.5})_1}{(t_{0.5})_2} = \frac{L_1^{1.55}}{L_2^{1.55}}$$

There is a definite difference in half-time for adsorption and desorption. As a rule of thumb, the half-time of adsorption is 25–30 % longer than desorption for the same wood species.

The change in temperature (ϑ) influences also the rate of moisture exchange. Because the temperature variation is generally not large, a simple linear conversion may be applied:

$$\frac{(t_{0.5})_1}{(t_{0.5})_2} = \frac{\vartheta_2}{\vartheta_1}$$

The kind of wood species certainly influences the rate of moisture exchange due to their differences in their internal structures. If reliable measurement results for a given species are not available, half-times taken from Fig. 2.6 may be used as a first approximation, naturally with the necessary conversions.

Moisture exchange with humid air can be considerably reduced using different surface treatments. A more demanding treatment is the soaking of wood in an appropriate monomer (e.g. diisocyanate) using vacuum impregnation (Németh 1986) or the WAN-drying and soaking method (Burmester and Olsen 1971). The latter method allows more uptake of monomers; therefore, the wood considerably decreases its hygroscopicity. The reason for reduction of sorption is the penetration of monomers into the cell wall shifting the sorption isotherm towards lower equilibrium moisture contents as shown in Fig. 2.7 (Burmester and Olsen 1971).The half-times also increase around 40 % compared to untreated wood. The reduction in hygroscopicity by impregnation also increases the dimensional stability of wood. The reduction in shrinkage in alternating temperatures may be as high as 50 %.

A more common and cheaper treatment of wood surfaces is painting them with different finishing materials.

Because moisture plays an important role in the photodegradation of wood, it is important to protect the wood surface from excess moisture uptake, especially outdoors. The moisture exchange through coatings will mainly be influenced by the permeability of the coating material to water vapour. The colour of a coating may significantly affect the reflection properties of the surface and thus on the surface temperature. The surface temperature of outdoor wood structures is governed by solar irradiation and to a lesser extent by convective heat transfer. The reflection of painted surfaces can change between the extremes from 5 % (black) to 85 % (white), and this highly affects the instantaneous surface temperature. A higher

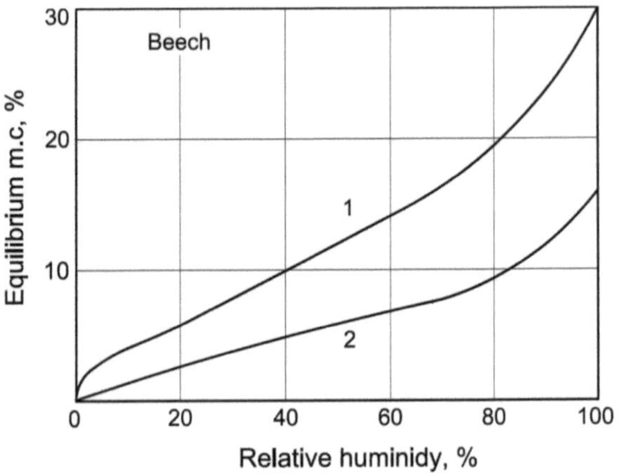

Fig. 2.7 Desorption isotherm of beech wood untreated (*1*) and soaked with diisocyanate (*2*)

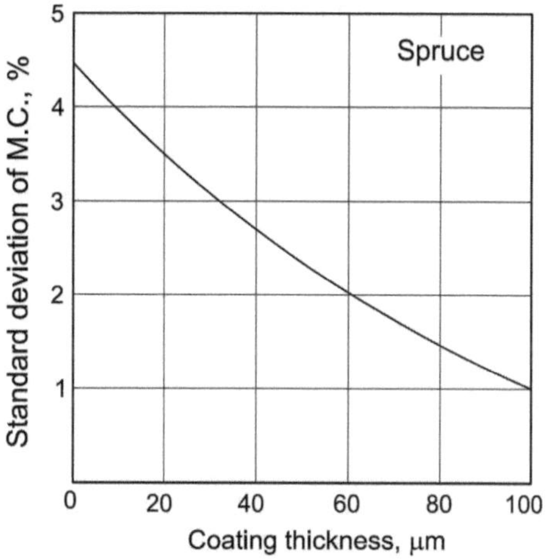

Fig. 2.8 Standard deviation of moisture content variation in a window frame as a function of coating thickness

temperature generates a higher vapour pressure inside the wood, causing a more intensive moisture exchange. This was verified experimentally as a black-painted window frame had less moisture content variations than a white-painted one (Sell 1985).

The permeability of coating generally depends on its thickness. Figure 2.8 shows the standard deviation of moisture content variation in a window frame exposed to weather as a function of paint thickness (Sell 1985).

2.4.2 Fluid Flow in Wood Materials, Infiltration

Wood is a porous material (because of its capillaries) always containing some water. Due to its hygroscopicity and permeability, wood has the ability to adsorb, conduct and lose water depending on circumstances.

The movement of moisture in wood can be described by diffusion-type equations. In many cases, the most common approach is to use the gradient of moisture content as a driving force in the form:

$$q = -D\frac{\partial W}{\partial x} \tag{2.13}$$

where D is the diffusion coefficient generally depending on the moisture content and W is the volumetric moisture content. The volumetric moisture content is related to the most commonly used moisture content by weight (X, dry basis) in the form given in the previous section.

In the soil science, the water potential is expressed in terms of energy per unit mass, J/kg, or in terms of unit volume (J/m^3 = N/m^2) giving pressure or pressure height.

The widely used pF number is the logarithm of the pressure height expressed in cm (1000 cm = 1 bar).

Using the water potential ψ (cm), Eq. (2.13) will be modified in the following way:

$$q = -D\frac{\partial W}{\partial \psi}\frac{\partial \psi}{\partial x} = -K(W)\frac{\partial \psi}{\partial x} \tag{2.14}$$

where

$$K(W) = D\frac{\partial W}{\partial \psi}$$

and it means the water conductivity coefficient (cm/h).

The water potential concept may equally be used for both bound and free water, and therefore, the gradient of water potential serves as the driving force within the whole range of moisture content.

The water potential as a function of moisture content is fundamentally bound to the pore size distribution in the wood material. The evaluation of the water potential curve occurs in the lower pressure range (at higher moisture contents) using the tension plate and pressure plate methods. The upper part of the water potential curve corresponds to the sorption isotherm because the relative humidity of air (φ) can be converted into tension:

$$\psi = \frac{RT}{m_{\mathrm{w}}}\ln\varphi \quad \mathrm{kJ/kg}$$

where

R is the universal gas constant (8.314 kJ/kmol K) and

m_{w} is the mass of the water molecule (18 kg/kmol).

Taking 50 % relative humidity, it corresponds to nearly 100 kJ/kg water potential and 1000 bar tension!

The evaluation of the water potential curve is a time-consuming procedure, and therefore, only a few studies on wood can be mentioned (Cloutier and Fortin 1991, 1993). Figure 2.9 shows the pore size distribution for a Pinewood as a function of pore radii (Sitkei 1994a). The corresponding theoretical water potential values are also given. The water potential curve of a Pinewood as a function of volumetric moisture content is given in Fig. 2.10.

It is interesting to note that the sorption isotherms for wheat corn and wood are very close to each other. Therefore, the upper part of the water potential curve is almost the same for both materials. The lower part of the curve highly dependent on porosity of the material, and therefore, different wood species, concerning the lower part of the water potential curve, may considerably differ from each other.

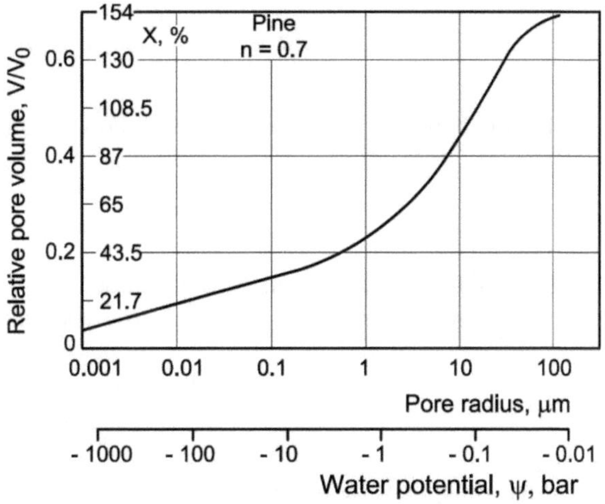

Fig. 2.9 Relative pore volume of a Pinewood as a function of pore radii. The corresponding theoretical water potential values are also given

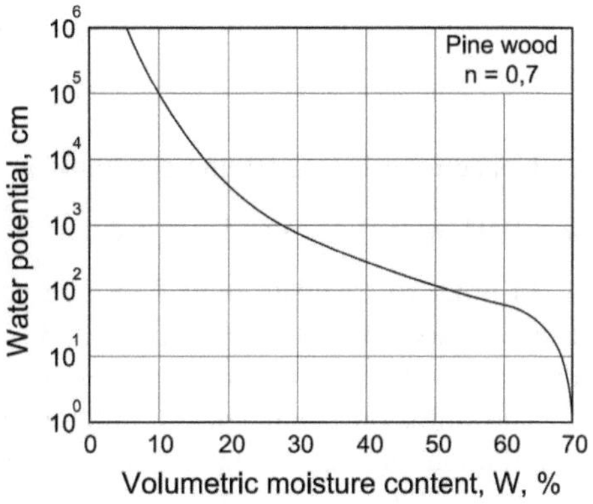

Fig. 2.10 Water potential of Pinewood as a function of volumetric moisture content

There are several mathematical expressions to describe the water potential curve. The exponential function is suitable for this purpose, except for very low tensions, in the following form:

$$\psi = -A\exp(-bW^n) \tag{2.15}$$

and

$$\frac{\partial W}{\partial \psi} = \frac{W^{1-n}}{\psi(W)nb} \tag{2.15a}$$

The exponent n has values around 0.1.

An other possible equation utilises the moisture ratio (MR) as follows (van Genuchten 1980):

$$\mathrm{MR} = \frac{W - W_r}{W_s - W_r} = [1 + (\alpha\psi)^n]^{-m} \tag{2.16}$$

so that the exponent m and n have the following interrelation:

$$m = 1 - \frac{1}{n}$$

The residual water content W_r is taken as 0.02–0.03.

Performing the derivation of Eq. (2.16) yields the following:

$$\frac{\partial W}{\partial \psi} = \frac{(W_s - W_r)mn\alpha^n\psi^{n-1}}{[1 + (\alpha\psi)^n]^{m+1}}$$

If the value of W_r is small compared to W_s and so it can be neglected, a simpler expression may be used:

$$\frac{\partial W}{\partial \psi} = \frac{W_s mn}{(\alpha\psi)^{mn}}$$

Concerning Fig. 2.10, the constants in Eq. (2.16) have the following values: $\alpha = 0.03$, $m = 0.2126$ and $n = 1.27$.

In order to calculate fluid flow, the conduction properties of the materials are needed. According to Eqs. (2.13) and (2.14), the conduction properties are the diffusion coefficient D and the water conductivity K(W).

The diffusion coefficient of different wood species depends on the pore size distribution of wood, the moisture content and the temperature. Measurement results for Pine and Beech are shown in Fig. 2.11 (Шубин 1965). The diffusion

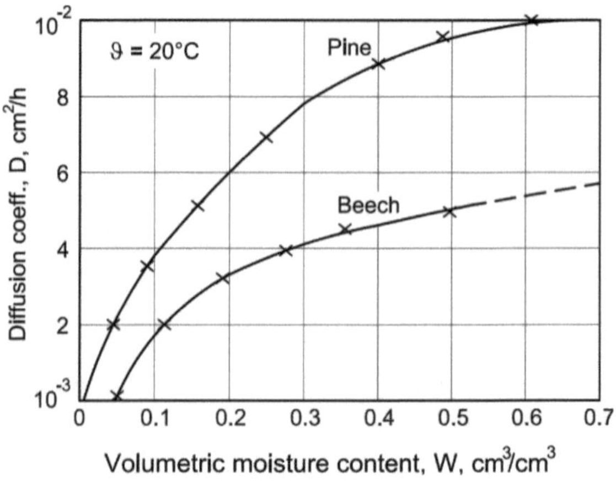

Fig. 2.11 The diffusion coefficients of Pine and Beech as function of volumetric moisture content

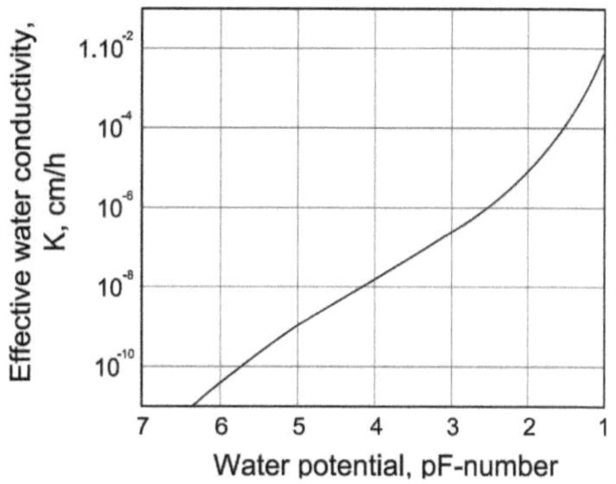

Fig. 2.12 Effective water conductivity in Pinewood as a function of water potential (tension)

coefficient increases with higher moisture contents and higher total pore volumes. An increase in temperature increases the diffusion coefficient considerably.

Knowing the diffusion coefficient and the derivative of the water potential, the water conductivity K(W) can be evaluated. Using the water potential of Pinewood in Fig. 2.10, the water conductivity as a function of the water potential is given in Fig. 2.12.

The water potential is fundamentally related to the pore size distribution of the given material. Some 50 years ago it was already demonstrated that the heat conduction coefficient for quite different soils could have been uniquely described as a function of moisture potential (Nakshabandi and Kohnke 1965). It may be supposed that the hydraulic conductivity K(W), as a first approximation, may have also general validity as a function of water potential, at least for the range $W < 50$ % (or $X < 100$ %). Depending on the overall porosity of the wood, differences may occur in the vicinity of the saturated state (K_s).

Putting water onto a wood surface, an infiltration process takes place. Water enters into the wood in response to the concentration or pressure difference, and the rate of entry is given by Eq. (2.13) or (2.14). These seemingly simple equations cannot be used because the gradients vary as a function of time and are not known. In order to examine the main characteristics of the infiltration process in time and space, the governing differential equation is needed.

For infiltration vertically downwards, the differential equation is written as follows:

$$\frac{\partial W}{\partial t} = \frac{\partial}{\partial y}\left(D\frac{\partial W}{\partial y}\right) \tag{2.17}$$

where y is the vertical coordinate, positive downwards from the surface. The problem described by Eq. (2.17) is that of diffusion of water into a wood column of infinite linear extent (semi-infinite body). Before starting the infiltration, the wood has initial moisture content W_{in}, and after starting the infiltration, the surface will become saturated (W_s).

In so far as the diffusion coefficient is moisture dependent (see Fig. 2.11), Eq. (2.17) can only be solved numerically. Taking a constant average diffusion coefficient for simplicity, however, Eq. (2.17) has analytical solution in the following form (Carslaw and Jäger 1959):

$$\frac{W - W_{\mathrm{in}}}{W_{\mathrm{s}} - W_{\mathrm{in}}} = erfc\left(\frac{y}{\sqrt{4Dt}}\right) \tag{2.18}$$

An inspection of Eq. (2.18) reveals that the solution is a function of $y/\sqrt{t}$. Such a combination of the two independent variables, well known in the diffusion theory, has some consequences:

- the propagation distance of the diffusing material specified by a given moisture content is proportional to $\sqrt{t}$;
- a given moisture content at any distance y from the boundary surface requires a time inversely proportional to D and proportional to y^2.

To demonstrate the propagation of moisture in a solid wood, calculations were performed (Fig. 2.13) showing the instantaneous water content profiles during infiltration in a Pinewood for different elapsed times.

The initial tangent of each curve gives the moisture gradient $\partial W/\partial y$ which would have been required to use Eq. (2.13).

The moisture gradient decreases with the elapsed time, and this means that the infiltration rate also decreases with time.

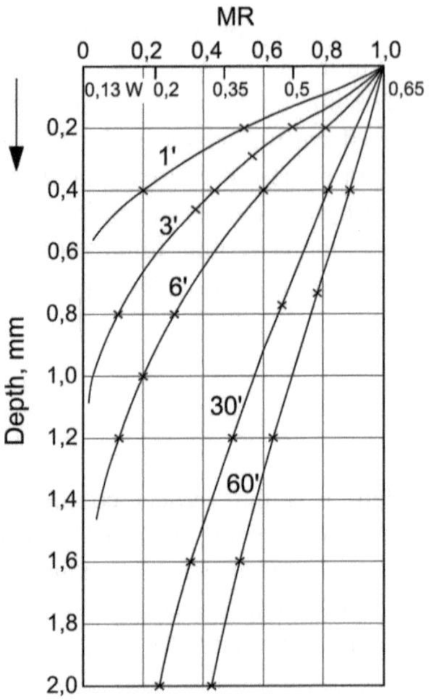

Fig. 2.13 Water content profiles during infiltration in Pinewood for different times. $D = 0.03\ \mathrm{cm^2/h}$, $W_{\mathrm{in}} = 0.13$, $W_{\mathrm{s}} - W_{\mathrm{in}} = 0.52$

The infiltration rate on the surface is given by the following equation:

$$q = (W_s - W_{in})\sqrt{\frac{D}{\pi t}} \quad \text{cm/h} \tag{2.19}$$

while the integration of this equation gives the cumulated infiltration for a given time:

$$Q = \frac{2}{\sqrt{\pi}}(W_s - W_{in})\sqrt{Dt} \tag{2.20}$$

Figure 2.14 shows the variation of surface flux and the cumulated infiltration of water in Pinewood as a function of elapsed time. The graph shows that a 1-mm layer of water requires *I* hour of infiltration time with the given diffusion coefficient.

Example: After 6-min elapsed time, the moisture gradient from Fig. 2.13 is $\Delta W/\Delta y = 0.52/0.1 = 5.2$ 1/cm and, using Eq. (2.13), $q = 0.03 \times 5.2 = 0.156$ cm/h = 1.56 mm/h. The exact value is 1.6 mm/h (see Fig. 2.14). If we would use Eq. (2.14), then we rewrite it into the following form:

$$q = K(W)\frac{\partial \psi}{\partial y} = K(W)\frac{\partial \psi}{\partial W}\frac{\partial W}{\partial y}$$

Keeping in mind that $W_{in} = 0.13$ corresponds to pF = 4.6 (see Fig. 2.10), therefore, we have

$$q = 3 \times 10^{-9}\frac{1}{10^{-7}}\frac{0.52}{0.1} = 0.156 \quad \text{cm/h}$$

which is the same result as obtained with Eq. (2.13).

From the above example, we conclude that the use of moisture gradient with Eq. (2.13) is simpler. But the use of moisture gradient as a driving force has the restriction that the material is homogeneous having the same water retention relationship in the entire space. It is well known in the practice that in a single stem, different moisture contents are in long-term equilibrium in the sap- and heartwood.

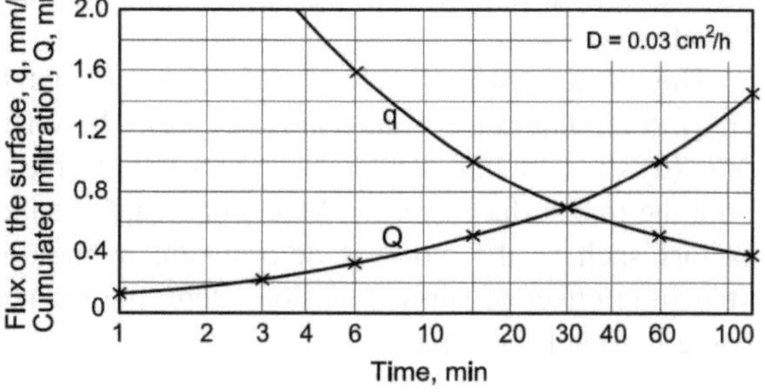

Fig. 2.14 Cumulated infiltration Q and water flux on the surface of Pinewood as a function of time

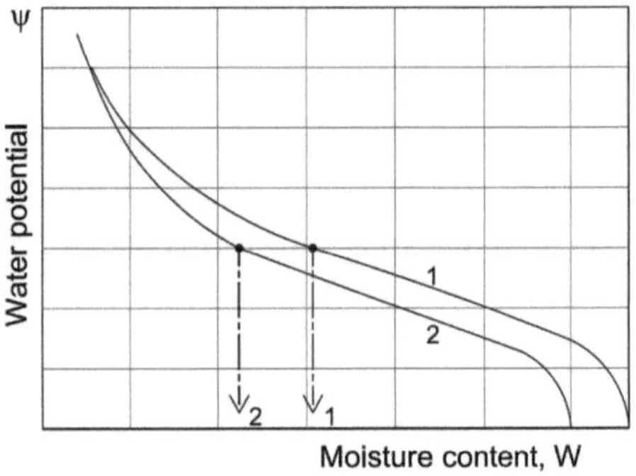

Fig. 2.15 Equilibrium of moisture in two different materials at a given moisture tension

The moisture gradient theory would say that this is simply impossible! The sap- and heartwoods have, however, different water retention properties (water potential curve), and using the water potential concept, equilibrium is possible at the same tension, but not at the same moisture content (see Fig. 2.15). It is obvious that the moisture gradient is not a true driving force in physical sense. At the same time, the space derivative of the water potential (energy) produces a force, thus corresponding to a real driving force. A further advantage is that the gradient in water potential can be used in the whole range of moisture contents.

In the possession of experimental results, and using Eq. (2.20), the average diffusion coefficient can simply and easily be estimated.

2.4.3 Surface Tension and Wetting Properties of Wood

At the boundary of a free liquid surface, the molecular forces are unbalanced and these inwardly directed forces create a new surface with *surface tension.* The molecules on the surface have greater potential energy than those in the interior, and the difference is the *free surface energy* of the liquid per unit area. The surface tension on the interface area between two liquids may equivalently be defined either through force or through energy. Surface tension is defined as force per unit length or as energy per unit area change. These two definitions are equivalent, but when referring to energy per unit of area, it is used the term "surface free energy" which is a more general term in the sense that it applies also to solid materials and not only liquids.

The surface tension is always related to two neighbouring media, i.e. liquid–air (γ_{LA}), solid–liquid (γ_{SL}) and solid–air (γ_{SA}). The surface tension has the dimensions N/m (dyn/cm) or J/m^2 (erg/cm^2), and for liquids, the surface tension may be equated with the free surface energy of the unit area. On solid surfaces, there are disturbing factors such as the work of deformation in the solid surface due to the lateral stretching effect of surface tension. Here, the surface tension and the free surface energy are not the same.

The surface tension acting on the boundary of the liquid tends to minimise the surface, and therefore, a small mass of liquid, where the gravitational force is small compared to the tension forces, will form a spherical convex body. Due to the surface tension, in the interior of the liquid a pressure is developed which is inversely proportional to the radius of the liquid surface (liquid is lying on the plane solid surface):

$$P_1 - P_2 = \gamma_{\mathrm{LA}} \frac{1}{R}$$

The effect of gravitational forces may only be neglected if the constant A characterising the ratio of surface tension and gravitational forces:

$$A = \sqrt{\frac{2\gamma_{\mathrm{LA}}}{\rho g}} \quad \mathrm{m}$$

is large enough compared to the size of liquid. Using water, $A = 3.8$ mm and in this case the gravitational forces up to 1-mm water thickness may be neglected. Indeed, for $R = 1$ mm, the pressure due to surface tension is 72 $\mathrm{N/m^2}$ and the hydrostatic pressure is only 10 $\mathrm{N/m^2}$.

Surface tension plays important role in the wetting process of wood surfaces. Two different kinds of wetting phenomena of practical interest may occur on wood surfaces:

- the gluing process at which the emulsion of a polymer adhesive first spreads on the solid surface and the measure of its wetting power is the contact angle,, but the joint itself depends on the adhesive work of the polymer system. It is termed as *adhesion wetting*;
- the spreading of liquid on a solid surface which is termed as *spreading wetting*.

The equilibrium of the surface tension of a liquid drop is illustrated in Fig. 2.16. The equilibrium in horizontal direction requires the following relation among the surface tensions (Young equation 1805):

$$\gamma_{\mathrm{LA}} \cos \Theta = \gamma_{\mathrm{SA}} - \gamma_{\mathrm{SL}}$$

or

$$\cos \Theta = \frac{\gamma_{\mathrm{SA}} - \gamma_{\mathrm{SL}}}{\gamma_{\mathrm{LA}}} \tag{2.21}$$

The contact angle may be smaller or larger than 90°, and its value depends on the relative magnitude of the different surface tensions. Therefore, the wetting or non-wetting condition is determined by the solid and liquid together.

The thermodynamic energy of interaction of two surfaces was formulated by Dupré (in 1869) as the work of adhesion:

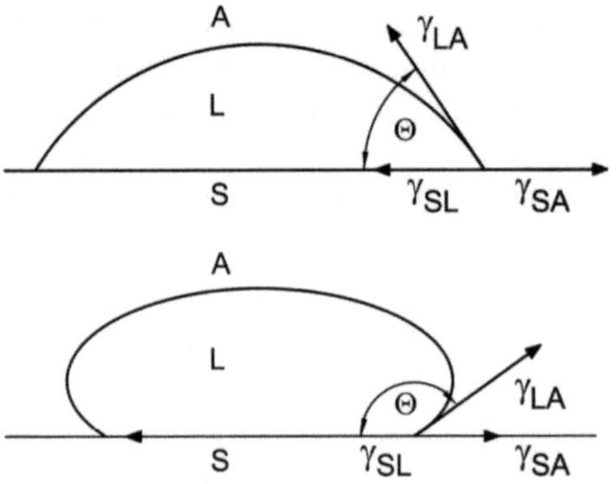

Fig. 2.16 The interfacial tensions between solid (*S*), liquid (*L*) and air (*A*) surfaces if the equilibrium contact angle (Θ) is between $0 < \Theta < 90°$ (*upper*) or between $90° < \Theta < 180°$ (*lower*)

$$-W_a = (\gamma_{SA} - \gamma_{SL}) + \gamma_{LA} \tag{2.22}$$

Similarly, the work of spreading is defined as follows:

$$W_s = (\gamma_{SL} - \gamma_{SA}) + \gamma_{LA} \tag{2.23}$$

From the above two equations, we get the following:

$$W_s - W_a = 2\gamma_{LA}$$

According to the definition, the adhesion work is always negative and the work of spreading is positive.

Combining Eqs. (2.21) and (2.22) yields the following:

$$-W_a = \gamma_{LA}(\cos\Theta + 1) \tag{2.22a}$$

$$W_s = \gamma_{LA}(1 - \cos\Theta) \tag{2.23a}$$

which are illustrated in Fig. 2.17. In theory, the contact angle $\Theta°$ may vary between zero and 180°, but in the practice, contact angles over 120° hardly ever occur. In Eqs. (2.22) and (2.23), the difference of two surface tensions, $\gamma_{SA} - \gamma_{SL}$, appears, and the measurement of the contact angle does not allow us to calculate the γ_{SA} and γ_{SL} surface tensions separately. An approximate estimation is, however, possible if we accept the validity of Antonov's rule.

Application of Antonov's rule to solid/liquid surface is based on the assumption that the interfacial tension (similar to that of between two liquids) is the difference between the surface tensions of the contacting solid and liquid. With this assumption

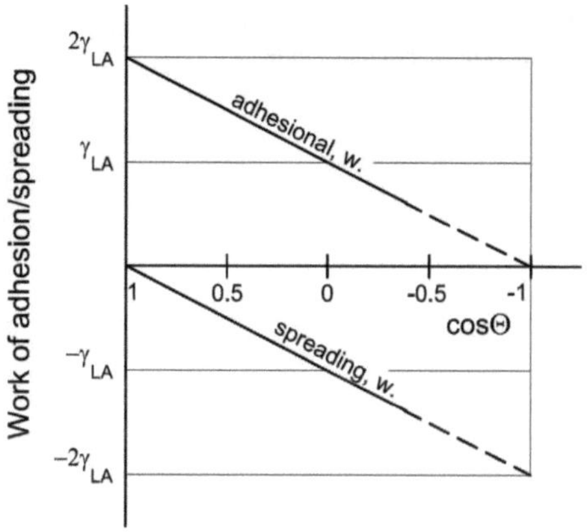

Fig. 2.17 Work of adhesion and spreading as a function of contact angle

$$\gamma_{SL} = \gamma_{LA} - \gamma_{SA}$$

and using Eq. (2.21) yields

$$\cos\Theta = \frac{2\gamma_{SA}}{\gamma_{LA}} - 1$$

and

$$\gamma_{SA} = \frac{\gamma_{LA}}{2}(\cos\Theta + 1) \tag{2.24}$$

Due to the application of Antonov's rule, the surface energy of adhesion and spreading may be expressed in the following simplified form:

$$-W_a + W_s = 2\gamma_{LA}$$

$$-W_a = 2(\gamma_{LA} - \gamma_{SL}) = 2\gamma_{SA} \tag{2.22b}$$

$$W_s = 2\gamma_{SL} \tag{2.23b}$$

Concerning wood surfaces and using water with γ_{LA} = 72 dyn/cm, the measured contact angles vary around an average of 75° (initial value), in this case γ_{SA} = 45 dyn/cm, γ_{SL} = 26 dyn/cm, and the surface energy for spreading wetting is 53 erg/cm^2 in good agreement with measured values. The surface energy for adhesion is 91 erg/cm^2. If the contact angles vary between 70° and 80°, the surface energy for spreading varies between 47.5 and 59.5 erg/cm^2.

In order to give a general overview, different measurement results were summarised and depicted in Fig. 2.18. The contact angle range of initial values for most

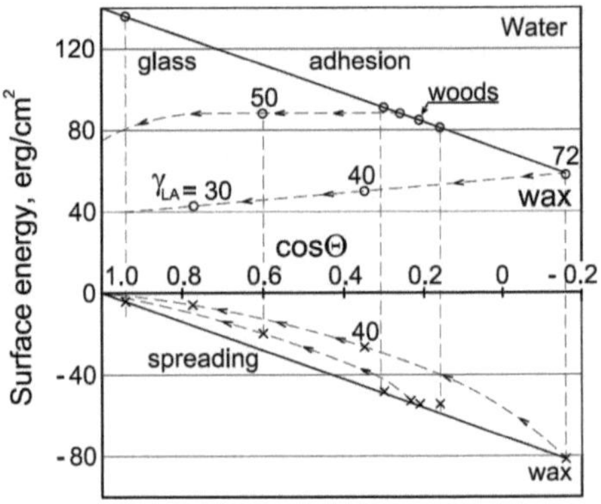

Fig. 2.18 Surface energies as a function of contact angle and surface tension

wood species is between 70° and 80°. After applying the liquid onto the surface, the contact angle rapidly decreases, due to infiltration and evaporation. Using an oil or wax surface treatment, the contact angle increases to near 90° and 102°–108°, respectively. The glass surface is given only to show the other extreme of the possible variations.

An interesting question is to substitute a liquid of lower surface tension instead of water (e.g. for petroleum γ_{LA} = 24 dyn/cm). The contact angle strongly decreases, but the work of adhesion remains more or less constant. The work of spreading decreases, however, considerably as indicated in Fig. 2.18 depending on the starting point.

Due to the complexity of the surface–liquid interaction, especially for wood surfaces, attempts were made to determine the expected various components of the total work of adhesion or spreading. It was claimed that the validity of Young's equation is limited to chemically homogeneous flat solid surface which does not adsorb liquid. The wood–liquid interaction is rather complicated because the surface of wood is rough, heterogeneous, porous, hydrophilic and anisotropic. That is why the equilibrium conditions of Young's equation are not exactly fulfilled for wood. The contact angle constantly changes over time, and this change implies problems in determining equilibrium. Liptakova and Kudela (1994) recommended: the equilibrium contact angle is defined as the contact angle measured in the moment of the change of the advancing contact angle to the receding contact angle.

The work of adhesion describes how the intermolecular forces contribute to adhesion between condensed phases. The total work of adhesion in an interfacial interaction between solids and liquids can be divided into two parts: the Lifshitz–van der Waals (LW) and the Lewis acid–base (AB) interactions (Fowkes 1983):

$$\gamma_a = \gamma^{LW} + \gamma^{AB} \tag{2.25}$$

The Lifshitz–van der Waals component can be calculated using nonpolar liquid as follows:

$$\gamma^{\mathrm{LW}} = 0.25\gamma_{\mathrm{LA}}(1 + \cos\Theta)^2 \tag{2.26}$$

The acid–base interactions include two different and independently variable properties expressed in the electron-acceptor (γ^+) and the electron-donor (γ^-) surface free energy parameters. This expansion leads to the "van Oss-Chaudhury-Good" model (Van Oss et al. 1987; Good 1993). The electron-acceptor (γ^+) and the electron-donor (γ^-) surface free energy parameters of wood can be calculated by applying polar liquids with known electron-acceptor and electron-donor parameters. The Young-Good-Girifalco-Fowkes equation can be applied using the measured contact angle data:

$$0.5(1 + \cos\Theta)\gamma_{\mathrm{L}} = \sqrt{\gamma_{\mathrm{S}}^{\mathrm{LW}}\gamma_{\mathrm{L}}^{\mathrm{LW}}} + \sqrt{\gamma_{\mathrm{S}}^{+}\gamma_{\mathrm{L}}^{-}} + \sqrt{\gamma_{\mathrm{S}}^{-}\gamma_{\mathrm{L}}^{+}} \tag{2.27}$$

This is the geometric-mean equation, and the harmonic mean equation can also be used (Gardner 1996). The Lewis acid–base component can be calculated as follows using the acceptor–donor components:

$$\gamma^{\mathrm{AB}} = 2\sqrt{\gamma^{+}\gamma^{-}} \tag{2.28}$$

The acid–base approach delivers the most detailed information about the surface chemistry, in giving values for acid and base components in addition to polar and disperse components of surface free energy, which is especially valuable with a chemically heterogeneous material like wood. However, it has to be stated that the results depend on the choice of test liquid. It is strongly recommended to work with minimum of three test liquids (Gindl et al. 2001).

The contact angle and surface free energy data of wood could be applied to predict the wettability of wood with different coatings and glues. The methods used to determine the surface energies of wood are generally based on static contact angle measurement of sessile drops or dynamic contact angle measurement [with Wilhelmy plate method (Gardner et al. 1991)] using the Young's equation (2.21).

Wood is a material with low free surface energy. This energy is additive. The free surface energy of wood is primarily composed of the Lifshitz–van der Waals component. Most of wood species show significant base component but only a very low acidic component. Table 2.1 presents the surface free energy parameters of some hardwood species (Gardner 1996). It is clearly visible that most of the total wood surface energy is the Lifshitz–van der Waals component. The acid–base component of the surface energy of wood is low. The base component is high enough, but the acidic component is small, almost zero. That is why the total acid–base component is small (see Eq. 2.28).

Table 2.1 Surface free energy parameters of some wood species (Gardner 1996)

Species	γ^{LW} (mJ/m^2)	γ^+ (mJ/m^2)	γ^- (mJ/m^2)	γ^{AB} (mJ/m^2)	W_s (mJ/m^2)
Ash (*Fraxinis Americana* L.)	42.63	0.001	67.35	0.52	43.15
Cherry (*Prunus serotina* Ehrh.)	47.46	0.42	28	6.84	54.3
Maple (*Acer saccharum* Marsh.)	45.48	0.46	33.19	7.85	53.3
Red oak (*Quercus rubra* L.)	39.67	0.46	37.74	8.33	48
White oak (*Quercus* spp.)	34.02	0.39	22.8	5.98	40
Walnut (*Juglans nigra* L.)	37.92	0.09	58.93	4.63	42.55

The surface energy of wood can be changed by surface modifications. Steaming is a proper treatment to modify the value of the surface energy. For this purpose, Black locust (*Robinia pseudoacacia* L.), Oak (*Quercus robur* L.), Merbau (*Intsia bijuga*) and Sapupira (*Hymenolobium petraeum*) samples were steamed at 92–122 °C (Varga and van der See 2008). During long-term, high-temperature steaming, Black locust and Merbau surfaces became strongly hydrophobic. However, contact angle between Oak and water was reduced by all of the applied steaming parameters (Fig. 2.19), which ensures a better wettability of the wood surface. Figure 2.19 shows that the value of the contact angle is decreasing rapidly right after the liquid contacts to the wood. After 2–3 s, the rapid decrease stopped and the contact angle decreased slowly.

The surface free energy of Black locust and Merbau slightly decreased during hydrothermal treatment, but there was no obvious relationship between the process parameters and the final energy value. The calculated energy values of treated Oak surfaces were higher than those of untreated samples. The surface free energy of Sapupira decreased slightly at low steaming temperatures. After long treatment at 122 °C, the surface energy of Sapupira increased significantly (Fig. 2.20). It can be concluded that the amount of surface free energy modified by steaming highly depends on the wood species.

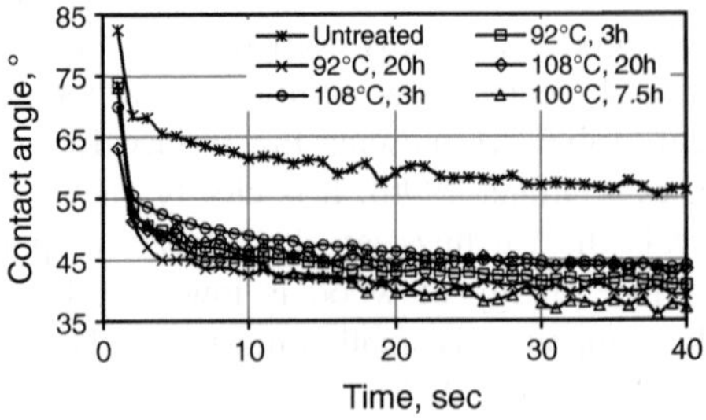

Fig. 2.19 Contact angle variation of steam-treated and untreated Oak as a function of elapsed time. *Test liquid* water (Varga and van der See 2008)

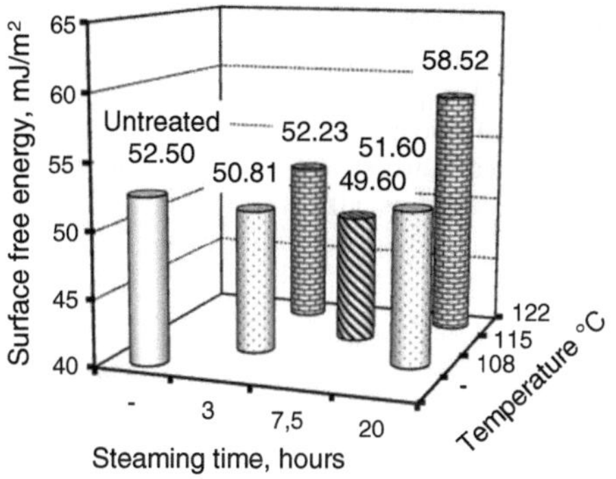

Fig. 2.20 Surface free energy of untreated and steam-treated Sapupira samples (Varga and van der See 2008)

Summarising the existing experimental results, we concluded that the interfacial interaction of wood surfaces with liquids has not been fully clarified in detail yet. We do not have a reliable relationship connecting the many variables derived from the pertinent properties of wood and liquids. Therefore, we need to seek simpler methods to characterise the spreading of a liquid on a solid surface, especially for practical purposes.

The spreading of liquids on a solid surface is the result of combined contribution of forces acting on the liquid. The surface tension of the liquid exerts a force perpendicular to the surface, and the components of this force in vertical and horizontal directions are highly dependent on the contact angle as it is shown in Fig. 2.21. For small contact angles, the horizontal components are also small enabling the horizontal spreading of the liquid. As a consequence, the thickness of the liquid layer on the solid surface will also be small. On the contrary, large horizontal forces resist spreading a liquid.

Measurements were conducted on various surfaces to determine the fluid thickness as a simple measure of spreading. Glass and variously treated wood surfaces were used, and water was applied onto the surface in different amounts. Measuring the mass and the diameter of the spreading liquid, the average thickness can easily be calculated. The results are depicted in Fig. 2.22. On the glass surface, thicknesses around 0.3 mm were measured, and on planed wood surfaces 0.7–0.8 mm were obtained, while on wax-treated surfaces the thickness was amounted to 1.9–2.0 mm. The results are in full agreement with contact angle measurements on paraffin, wax and glass giving extreme high and low contact angles, respectively (Garrett 1964).

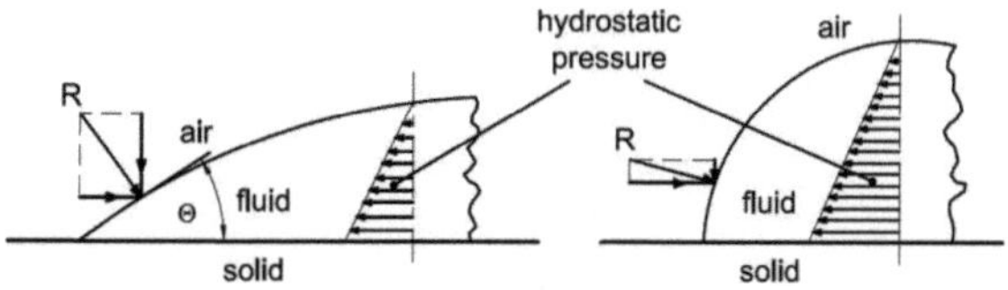

Fig. 2.21 Forces on a spreading fluid with different contact angles

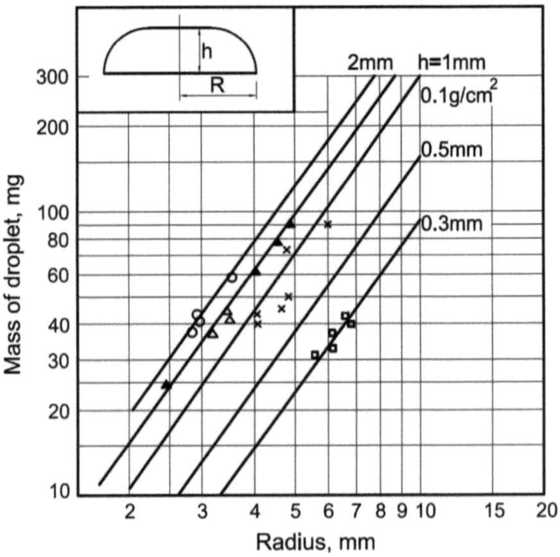

Fig. 2.22 Thickness evaluation on various surfaces. (□—glass; ×—planed wood; Δ—oil-treated wood; o—wax-treated wood)

A further proof for a possible interrelation between thickness and contact angle is given in Fig. 2.23, which represents the measured data as a function of cos Θ. The obtained relationship is a straight line (with good approximation) which corresponds to the theoretical curve according to Eq. (2.23).

Practical experience shows that thicker paint does not spread as well as thin paints, which is mainly due to the difference in their viscosities. At the same time, the effect of viscosity is never treated in spreading and contact angle problems. To study the possible influence of viscosity, measurements were carried out on the same surfaces as formerly, but using low fat (10 %) milk cream with a surface tension of 48 dyn/cm (65 % compared to water) and dynamic viscosity of 2.0 Ns/m^2 which is twice as high as that of water.

If viscosity had no influence on spreading, then a smaller surface tension should give better spreading. On a glass surface, the average thickness was 0.42 mm

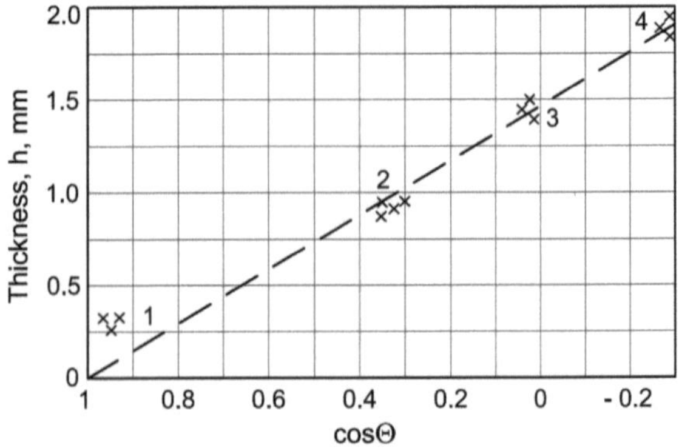

Fig. 2.23 Relationship between the average layer thickness and contact angle. *1*—glass; *2*—planed wood; *3*—oil-treated wood; *4*—wax-treated wood

(0.3 mm for water), and on planed wood surfaces, it was 1.0 mm (0.7–0.8 mm for water). On treated wood surfaces, the thickness was much the same as for water. From these results, we concluded that spreading is influenced by the viscosity to a given extent. Surface tension and viscosity have no definite connection with each other. For example, heating water from 20 to 100 °C, the viscosity decreases to 30 % and the surface tension only to 80 % compared to their initial values.

The following principles should be used in the adhesive joint process: A low surface tension γ_{LA} of the adhesive is preferable when applying adhesive onto a surface. The spreading is enhanced with high surface tension γ_{SA} and a low value of γ_{SL}. The assembly of the joint pairs, keeping in mind Eq. (2.22) for work of adhesion, will be enhanced by high values of γ_{SA} and γ_{LA}, and low value of γ_{SL}.

The optimum energy conditions for the overall adhesive joint process may be summarised as follows. The solid surface tension γ_{SA} should be as high as possible. The surface–liquid tension γ_{SL} should be as low as possible. Latter is equal to the statement that the adherend would possess affinity or interaction with the adherent. If we remember now the simplified Eqs. (2.22b) and (2.23b), we see that the energy of adhesion mostly depends even on γ_{SA} and the energy of spreading depends on γ_{SL}. The optimum surface tension of adhesive γ_{LA} may depend on different requirements. A good spreading requires lower values, while high joint strength needs higher surface tension. Furthermore, other factors such as the use of solvents and additives may influence or limit the optimum choice of surface tension γ_{LA}.

2.5 Photodegradation of Wood

2.5.1 Measurement Techniques for Studying Wood Degradation

The basic interaction during light and material contact is the interaction between one photon and one electron. The result of the interaction is highly determined by the energy (E) of the photon. This energy is in proportion to the frequency (v) of the proton.

$$E = hv \tag{2.29}$$

where h is the Planck constant ($h = 6.626 \times 10^{-34}$ Js). The UV photons of sunlight have enough high energy to break many of bonds in chemical components of wood. These alterations change the chemical and physical properties of the surface layer. These changes can be monitored by chemical and physical measurement techniques.

The most commonly used measurement techniques are the colour measurement, the infrared (IR) spectrum determination, and the roughness measurement. These tree techniques are discussed in this chapter. The electron microscopy is also a proper method in some special cases, and further chemical methods are available to clarify the details.

The degradation of wood appears mostly on the surface. The most sensitive surface parameter is the colour. The colour is mainly determined by the extractive content of wood. The colour change is created by the alteration of conjugated double-bond chemical systems. These systems are present originally in the lignin and in the extractives of wood. The degradation can split the chemical bonds followed by the oxidation of the radicals that were created. These processes can also create conjugated double-bond chemical systems. Typical example is the thermal degradation of hemicelluloses producing brown colour.

The objective colour measurement requires some rules. The colour of an object is highly determined by the emission spectra of the applied light source. The most preferable light source is the sun. The emission spectrum of the sun is changing slightly. That is why the International Commission on Illumination (CIE) defined the D_{65} standard illuminant. The D_{65} is intended to represent average daylight and has a correlated colour temperature of approximately 6500 K. The colour slightly depends on the direction from which it is observed. That is why the angle between illumination and observation must be the same for comparison. The most common angle is 10° for colour measurement of wood. The colour appearance of wood is usually inhomogeneous. Large test-window application is recommended for getting the average colour of the sample. There are available many systems to evaluate the colour and colour change of wood. The techniques of colour measurement for characterisation of photodegradation are described in Sects. 3.2 and 3.3.

The mostly used system is the CIELAB, three-dimensional rectangular coordinate system. This system is standardised in most countries. The L^*, a^* and b^* are the Descartes coordinates of the colour space in the CIELAB system (see more details in Sect. 3.2). The a^* and b^* coordinates are useful parameters to follow the chemical changes separately. However, for monitoring the visible colour change, the b^*/a^* quotient is the advantageous parameter.

The total colour change (ΔE^*), as a single number, is also often used. It represents the distance between two colour dots in the 3-dimensional colour space. It is usually applied to represent the colour change by one value. The ΔE^* is, however, not always reliable by itself. In some cases, ΔE^* values do not match to the visual observations. The individual colour coordinates (L^*, a^*, b^*, C^* and h) are more suitable to monitor the colour change caused by the photodegradation than the total colour change (see more information in Sect. 2.6.3).

Several papers try to present correlation between the total colour change and individual chemical changes. The real problem is that these researchers believe that they have found real correlation. However, total colour change is an extremely complex parameter since it is determined by the alterations of all three colour coordinates. Nevertheless, these colour coordinates are calculated by extremely complicated equations applying the tristimulus values. Furthermore, a^* and b^* encompass two independent tristimulus values calculated by integrals based on the reflection spectrum. (These integrals are not presented in this work.) The reflection spectrum carries information on the chemical changes through the absorption change of visible light. It is clear that the absorption change values undergo several mathematical manipulations to obtain the total colour change. As wood is a chemically complex material, absorption changes are never created by a single chemical change. Thus, the total colour change is a result of extremely mixed individual absorption changes.

Finding correlation between the total colour change and increase of the IR absorption of the carbonyl groups during photodegradation is a common mistake. Nevertheless, authors can be happy because they are misled by the high value of the calculated coefficient of determination (R^2). Unfortunately, it is not well known that the high value of R^2 does not necessarily mean that the correlation exists. It only indicates how strong it is—providing the correlation exists at all. The origin of the mistake is that there is similar exponential time dependence of the total colour change and that of the absorption increase for carbonyl groups. However, it is not sufficient condition. A true correlation requires that the two phenomena are also physically connected to each other.

This artefact can be visualised easily. If someone measures the velocity of free fall body for some seconds and measures the same of an accelerating car and examines the correlation between the measured data, will be surprised. Calculating the R^2 value, it will be almost one unit. Hopefully, no one think that the free fall is determined by the accelerating car.

The change of surface roughness (see in Sect. 4.4 for details) is also a proper indicator to monitor the degradation of the surface. The roughness change is not as sensitive as the colour change, especially in case of short-term degradation.

The optical roughness measurement methods are usually more susceptible than the direct contact methods, but these are highly sensitive to the size of the measured area. A newly developed method utilises the fact that the light scattering depends on roughness. The increasing roughness produces greater scattering. The diffuse reflection spectrum of a sample integrates the absorption and the scattering as well. The absorption is characteristically wavelength dependent, while the scattering is slightly wavelength dependent and this correlation is close to linear. Looking at the reflection spectrum, the negative peaks represent the absorption in opposite format and the scattering produces the so-called baseline shift (Fig. 2.24). Quantifying the baseline shift, data can be created which correlates to the roughness. The baseline shift can be easily determined if there are minimum two wavenumbers (or wavelengths) where there are no absorption. The IR spectrum is usually presented as the function of wavenumber instead of wavelength. The wavenumber ($1/\lambda$) represents the number of waves located within one centimetre, and it is determined as follows:

$$\frac{1}{\lambda} = \frac{v}{c} \quad 1/\mathrm{cm} \tag{2.30}$$

where

λ is the wavelength,
v is the frequency and
c is the velocity of the light.

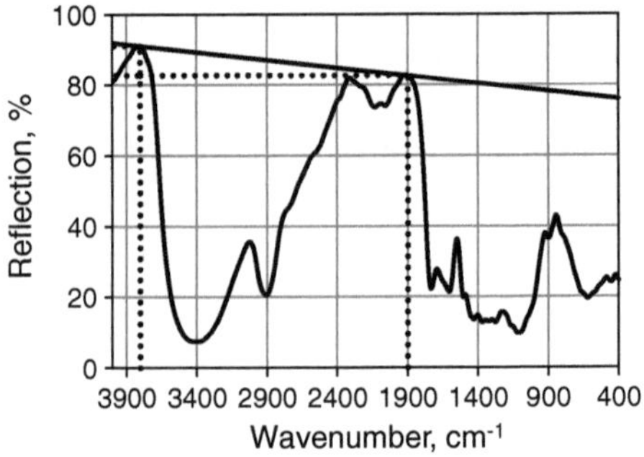

Fig. 2.24 The diffuse reflection infrared spectrum of Spruce wood indicating the baseline shift

Wood has no absorption at 1900 and 3800 cm^{-1} wavenumbers. It would mean the reflection should be 100 % at these places. But the reflection is less than 100 % because of the light scattering; this is called the baseline shift in IR spectroscopy (Fig. 2.24). Calculating the absorption spectrum, these reflection values are lifted up to 100 % (lifting up the whole spectrum) to eliminate the scattering effect. This is the baseline correction, which is regularly used in diffuse reflectance IR spectroscopy. The scattering depends on the roughness of the surface, and therefore, the baseline shift gives information about the surface roughness. Here, the baseline shift is given as the quotient of two reflection intensities at 3800 and 1900 cm^{-1} (Tolvaj et al. 2014a).

$$\text{Baseline shift } = R(3800)/R(1900) \tag{2.31}$$

where $R(3800)$ is the reflection intensity at 3800 cm^{-1} and $R(1900)$ is the same at 1900 cm^{-1}.

The IR reflection spectra of several wood species were recorded. After calculating the baseline shift parameters (2.31), the correlation between roughness and baseline shift was evaluated. The results are presented in Fig. 2.25 for Black locust and Scots pine species. The linear trend lines are parallel and are located close to each other. The high values of the R^2 show good correlation between roughness and baseline shift in all cases. This close correlation suggests that it is possible to

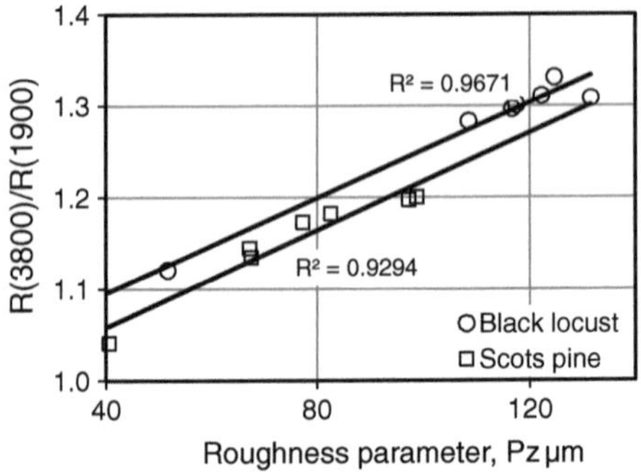

Fig. 2.25 Correlation between baseline shift and roughness of Black locust and Scots pine samples

monitor the roughness alteration effect of photodegradation by measuring the IR diffuse reflection infrared spectra and calculating the baseline shift (Eq. 2.31). It is highly important to note that exactly the same area of the sample must be used for all reflectance measurements; otherwise, the surface inhomogeneity overlaps the roughness change.

The infrared spectrum measurement is a widely used chemical method for determining the changes caused by different treatments. Moreover, it is a useful technique to determine the chemical composition of a material. The IR portion of the electromagnetic spectrum can be divided into three regions: the near-, middle- and far-infrared regions. These are named based on their relation to the visible spectrum. The middle region or analytical region, approximately 4000–400 cm^{-1} (2.5–25 μm), may be used to study the basic vibrations and associated rotational–vibrational structures. This region is the most frequently used for chemical analysis. Infrared spectroscopy utilises the fact that molecules absorb specific frequencies that are characteristic of their structure. The so-called fingerprint region is located within this analytical wavenumber range. Reflectance techniques are used mainly to measure the IR spectrum of wood. It is necessary to measure both the sample and a reference to take the infrared spectrum of a sample. As a usual measurement method, first the reference (background spectrum) must be measured and then replaced by the specimen, and the spectrum of the sample is measured. The proper material for the reference measurement must have bad mirror reflection and excellent diffuse reflectance. Instead of a good metal mirror, a rough surface is recommended with particles as small as possible. KBr powder is often used for this purpose. The KBr powder must be freshly dried, because this powder is strong water absorber. Water has many absorption peaks in the analytical IR region, and therefore, air humidity must be kept constant within the spectrophotometer and the sample compartment. The absorption spectrum is calculated in comparison with the reflected intensity data of the background material and that of the sample. High intensities in the background spectrum generate high intensities in the reflectance spectrum of the sample too. That is why the proper, good quality background spectrum is important for getting high-quality reflectance spectrum. It means that the reflection intensity data are repeatable only if the same material is used to measure the background spectrum (Tolvaj and Mitsui 2011). Fourier-transform infrared (FTIR) spectroscopy is used as a powerful technique nowadays. The equipment produces an interferogram by the interferometer. Then, the interferogram is Fourier-transformed to get the actual spectrum.

Since wood is not transparent for light, therefore, the traditional technique (measuring the light intensity before and after the sample) cannot be used to determine its absorption spectrum. Before creating the powerful FTIR technique, the only possibility was to pulp the non-transparent sample and press a small amount of the sample into KBr pellet. This pellet was transparent for IR light to determine the absorption spectrum of the non-transparent sample. The FTIR technique makes it possible to measure the diffuse reflectance spectrum. The reflectance spectrum contains the absorption and the scattering as well. Using the Kubelka–Munk (K-M) theory, the quotient of the absorption and scattering as the function of the wavenumber can be calculated. The diffuse reflectance infrared Fourier-transformed (DRIFT) spectrum provides information about a thin surface layer. Using this method, the diffusely reflected light is collected, and the absorption is calculated by the Kubelka–Munk equation.

$$\frac{k}{s} = \frac{(1 - R)^2}{2R} \tag{2.32}$$

where k, s and R are the absorption coefficient, the scattering coefficient and the reflectance of a thick enough layer, respectively. Fortunately, the scattering coefficient hardly depends on the wavenumber of the IR light, so the shape of K-M function is similar to the wavenumber dependence of the absorption coefficient. This K-M function is usually used as the IR absorption spectrum. The K-M theory is recommended for poorly absorbing materials. As wood is a strong IR light absorber in some wavenumber region (e.g. the absorption region of ether bands 1100–1200 cm^{-1}), the usage of K-M equation has limits to its validity (details will be discussed later).

Measuring the IR spectrum is a commonly used method to monitor the chemical changes of wood caused by different treatments. The treatment usually changes the roughness of the surface, and the increase in roughness results in the increase of light scattering. This phenomenon creates baseline shift. In this case, baseline correction must be done. The baseline correction is recommended to be done for reflectance spectrum before K-M transformation. There are two-point, three-point and multi-point baseline corrections. The computer supported IR spectrophotometers does the baseline correction automatically if the proper points are chosen. Basic points can be those wavenumbers where the sample does not have absorption. Neither of the wood species have any absorption at 1900 and 3800 cm^{-1} wavenumbers which means that two-point baseline correction can be applied for wood in all cases. Selecting a wavenumber with absorption, the spectrum may suffer dangerous alteration creating artificial changes!

Figure 2.26 shows the measured diffuse reflection spectrum of an Oak sample. The intensity of reflection should be 100 % at 1900 and 3800 cm^{-1} wavenumbers, but these intensities are less than 100 % because of the scattering apparently.

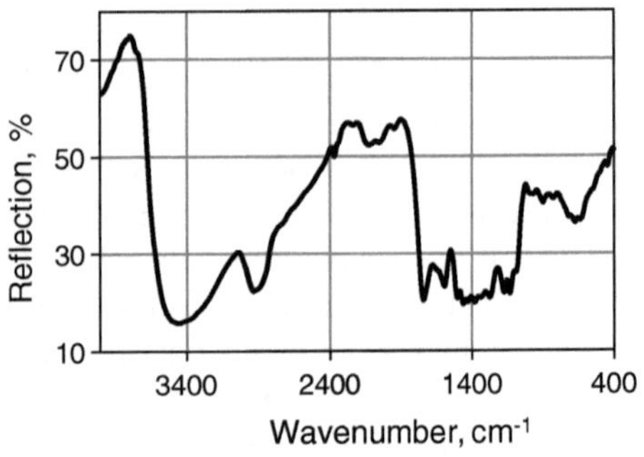

Fig. 2.26 The diffuse reflection spectrum of an Oak sample

Applying two-point baseline correction at 1900 and 3800 cm^{-1} wavenumbers, the spectrum of Fig. 2.26 will be converted as it is seen in Fig. 2.27. This process mostly eliminates the effect of scattering. This elimination is not perfect, however, because the wavenumber dependence of scattering is not precisely linear.

The K-M transformation (Eq. 2.32) converts the reflection spectrum into K-M spectrum which is similar to the absorption spectrum. The K-M conversion of Fig. 2.27 can be seen in Fig. 2.28 for Oak.

The left half of the spectrum consists of two broad bands. The higher peak belongs to the absorption of hydroxyl groups. Hydroxyl groups are located at various places in cellulose, hemicellulose and lignin. These diverse OH groups have absorption at different wavenumbers causing a rather wide absorption band. Water in the wood also results in absorption in this region. The second band represents the absorption of different methyl groups. The chemists created a list of absorption peaks of IR spectrum for identifying the absorbing chemical groups of the measured sample. This list is presented in Table 2.2 with the assignment of the bands. The list represents the maxima of the peaks. The place of the maxima slightly depends on the wood species. That is why intervals are given in Table 2.2. The intervals were determined by experimental data. Since the fingerprint region (between 800 and 1800 cm^{-1}) consists of several overlapping bands, some of the peak maxima are not

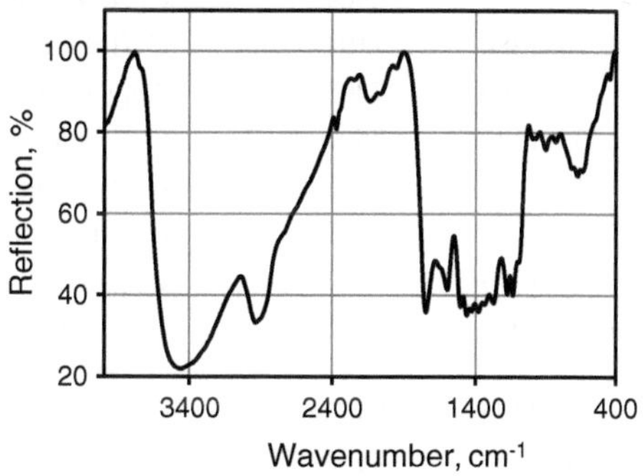

Fig. 2.27 The baseline corrected diffuse reflection spectrum of an Oak sample

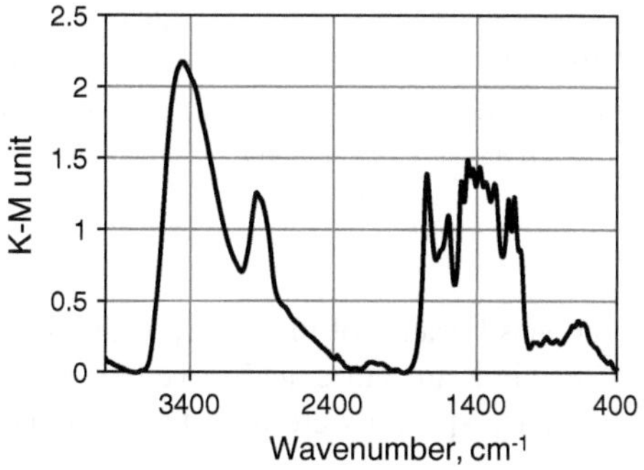

Fig. 2.28 The K-M spectrum of an Oak sample

Table 2.2 Characteristic IR bands of wood (place of maximum) and band assignments (Tolvaj and Faix 1995; Calienno et al. 2014; Huang et al. 2008)

Wavenumber (cm^{-1})	Wavelength (nm)	Assignment
3600–3550	2778–2817	Intramolecular hydrogen bond in a phenolic group (in lignin) and weakly bounded absorbed water
3360–3310	2976–3021	O(3)H..O(5) intramolecular hydrogen bonds in cellulose
2928	3415	CH_2 stretching asymmetric
2854	3504	Symmetric CH_2 stretching
1770–1757	5650–5692	C=O stretching vibration of non-conjugated ketones and γ lactones
1736–1705	5760–5865	C=O stretching vibration of carboxyl groups and acetyl groups in hemicelluloses (xyloglucan)
1660–1653	6024–6050	Conjugated C–O in quinines coupled with C=O stretching of various groups (flavones)
1628–1618	6143–6180	C=O stretching in flavones
1604–1594	6234–6274	Aromatic skeletal breathing with CO stretching (syringil lignin)
1512–1505	6614–6645	Aromatic skeletal (guaiacyl lignin)
1478–1476	6766–6775	C–H deformation in lignin
1465–1457	6826–6863	C–H deformation in xylan
1435	6969	C–H deformation in lignin and carbohydrates
1390–1380	7194–7246	C–H deformation in cellulose and hemicellulose
1369–1366	7305–7321	Aliphatic C–H stretching in methyl and phenol OH
1333–1342	7502–7452	C–H deformation, C-OH stretching, syringyl ring
1319	7582	C–H_2 wagging, C–H deformation (conifers)
1285	7782	C–H bending mode in cellulose
1285–1275	7782–7843	C_{aryl}-O, guaiacyl ring breathing with CO stretching
1240–1230	8065–8130	C–O linkage in guaiacyl aromatic methoxyl groups and acetyl groups in xyloglucan
1183–1175	8453–8511	C–O–C stretching (asymm.) in cellulose and hemicelluloses
1158–1156	8636–8726	C–O–C stretching in pyranose rings, C=O stretching in aliphatic groups
1138–1131	8787–8842	C–O–C stretching (symm.), arom. C–H i.p. deformation, glucose ring vibration
1108–1106	9025–9042	C–O–C stretching
1078–1076	9276–9294	C–O stretching mainly from C(3)–O(3)H in cellulose I
1050–1045	9524–9569	C–O and C–C stretching in cellulose and hemicelluloses
898	11,136	C–H deformation of cellulose

real ones because of the superposition of two or more individual bands. The places of maxima are not exactly overlap with the maxima in Fig. 2.28. Therefore, the difference spectra were used to determine the places of maxima.

The "fingerprint" region is about 1000 cm^{-1} wide interval and slightly depends on the complexity of the material. The usual interval is between 500 and 1500 cm^{-1}. In case of wood, the fingerprint region is located between 800 and 1800 cm^{-1}. The difference spectrum method helps to identify the changes. After creating the difference spectrum (irradiated minus initial), only those absorption bands appear where changes occurred. The increase in absorption is represented by positive band, while negative band represents the absorption decrease. The differences are usually small compared to the intensities of the original spectrum. That is why the disturbing effects must be minimalized. These effects for wood are inhomogeneity and anisotropy of the sample and the roughness change during treatment. The same area of the sample must be measured before and after the treatment. The grain direction, the angle between the fibre direction and the incident IR light must be identical for all measurement (Zavarin et al. 1990; Tolvaj and Mitsui 2004).

The increase of the surface roughness during treatment may cause difficulty if the reflection intensity is low. It happens if the absorption or the scattering are strong. The K-M equation determines the quotient of absorption and scattering. This quotient is something like the hyperbolic function of the reflection intensity (Eq. 2.32). The rough surface elevates the light scattering and decreases the collected light intensity. The K-M intensities increase enormously due to the increasing scattering and not because of increase in absorption (Tolvaj and Mitsui 2011). This phenomenon is visible in Fig. 2.29. The increase at 1740 cm^{-1} is not surprising since it is the absorption place of non-conjugated carbonyl groups and the number of these groups is increasing during UV irradiation. However, most part of the fingerprint region reflects the lifting effect of the increasing scattering. The low absorption interval between 1450 and 1650 cm^{-1} is not affected by the lifting and the absorption decrease at 1507 cm^{-1} is well visible. The lifted lines contain the absorption changes as well, but it is difficult to find them.

The difference spectra clearly show the superposition of the absorption change and the lifting effect (Fig. 2.30). The usual changes are visible after one-day treatment. There are clear absorption decreases at 1600 and 1508 cm^{-1} indicating the degradation of lignin. The ether band splitting is also visible between 1130 and 1180 cm^{-1}. Two positive peaks emerged at 1705 and 1763 cm^{-1}. A small lifting

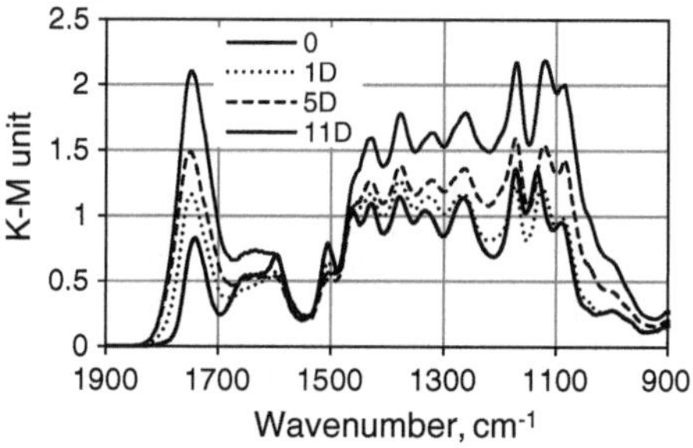

Fig. 2.29 The K-M (absorption) spectra of Beech after 1, 5 and 11 days (D) of UV irradiation

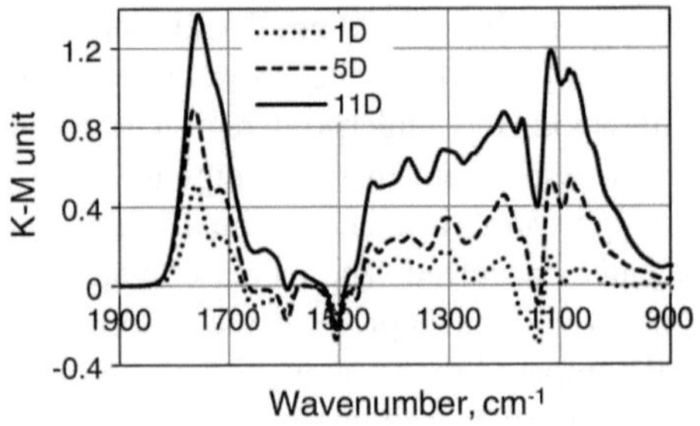

Fig. 2.30 The difference spectra of Beech after 1, 5 and 11 days (D) of UV irradiation without normalisation

effect is visible between 1000 and 1450 cm^{-1} caused by 1-day treatment. The increasing treatment time amplifies the lifting effect, and it becomes more and more visible. After 11 days of irradiation, almost the whole difference spectrum was lifted up. The negative peaks of ether bond degradation seem to be a valley between two positive peaks after 11 days of irradiation.

In similar cases, the spectrum normalisation can help to obtain a more correct difference spectrum. The normalisation can be applied properly if there is a peak not affected by the applied treatment. The band maximum around 1375 cm^{-1} can be the proper normalisation peak for wood during the photodegradation. This C-H band of cellulose is often used as internal standard because of its high intensity and central position. This peak can only be used for normalisation if the examined treatment does not influence this cellulose band. During the normalisation process, the absorption intensities of the treated sample are multiplied to get the same intensity at the normalising place as the spectrum of the untreated sample. The other possibility is to multiply both spectra to get the unity (1.0) at the normalising place. This second normalisation can be recommended if more than two changes are compared. The difference spectra of Fig. 2.29 are seen in Fig. 2.31 after normalisation at 1375 cm^{-1}. Most parts of these difference spectra represent the absorption changes correctly. A moderate lifting effect is visible only between 900 and 1320 cm^{-1}.

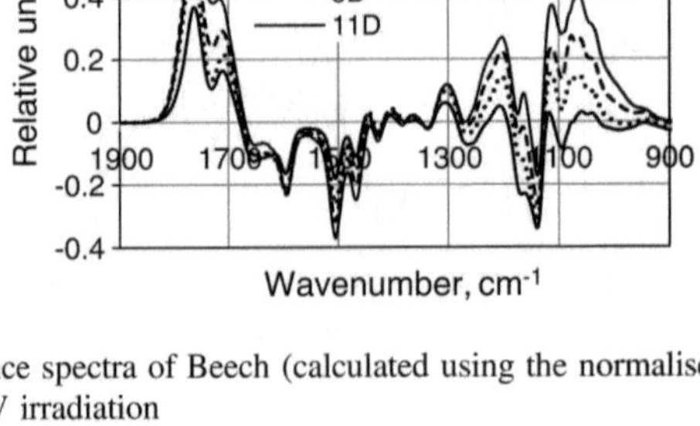

Fig. 2.31 The difference spectra of Beech (calculated using the normalised spectra) after 1, 3, 5 and 11 days (D) of UV irradiation

Comparing Figs. 2.30 and 2.31, the beneficial effect of normalisation is obvious. The remained lifting effect must be taken into consideration during the difference spectrum evaluation in this highly absorbing region.

Beside the diffuse reflectance technique, the IR absorption spectrum of non-transparent materials can be determined with the attenuated total reflectance (ATR) method. The total reflection works if the crystal is made of an optical material with a higher refractive index than the sample being studied. In that case, the light does not cross deeply the boundary of the crystal but will be totally reflected if the incident angle is large enough. During this reflection, the light slightly penetrates the second material (examined sample) and suffers minor absorption. There must be close contact between the crystal and the sample. The sample surface must be as flat as possible, and it must be pressed to the surface of the crystal to ensure as strong contact as possible. The ATR measurement is only partly applicable for the examination of wood because of its high porosity. It follows that the real contact area between the crystal and the wood material is always smaller than the actual surface despite of the press. Most of the applied crystals could not tolerate the pressure that the wood would require, except the diamond, but it shows strong absorption in the absorption region of non-conjugated carbonyl groups, highly reducing the sensitivity in this region.

The third possible method to determine the IR spectrum of wood is the photoacoustic method. The sample is irradiated by IR light rapidly interrupted with a rotating slotted disc. The absorbed energy from the light causes the local heating of the sample, and the generated thermal expansion of the surrounding gas produces pressure wave. This pressure wave is detected by a microphone. The wave intensity correlates to the absorbed energy, and the computer creates the absorption spectrum of the sample using the amplified microphone signal. This method does not need any special sample preparation.

2.5.2 Mechanism of Photodegradation

As the energy of photon increases proportionately with its frequency, the blue light and the UV radiation can be effective to produce degradation. Photodegradation is usually the first step for decomposition of outdoor wooden applications. The UV part of the sun radiation splits some chemical bonds producing free radicals due to the dehydrogenation, dehydroxylation and demethoxylation and generates chain scissions in cellulose and hemicellulose. The free radicals react with oxygen forming hydroperoxides. The lignin and the extractives are good light absorbers, and these components suffer the major degradation during UV radiation. The consequence of this photodegradation is embodied in drastic changes in the appearance of wood showing discolouration, lightness decrease (darkening), loss of gloss, roughening and checking of the surface. Then, the rain can leach out the products of degradation, making room for further degradation. These processes cause the destruction of mechanical and physical properties of the surface layer.

Wood is not transparent for light because of the wide range of chromophoric groups or systems distributed in surface components of wood. The measured depth of light penetration depends partly on the measuring method. The infrared absorption change can be used as an indicator. The absorption decrease directly shows the cleavage of chemical bounds caused by light absorption. The absorption increase indicates the newly created chemical groups. This is a secondary process, and in most cases, the degradation process is followed by oxidation. It may affect deeper layer than that reached by light. Kataoka et al. (2005) found that there is an inversely proportional relationship between the maximum depth of photodegradation and wood density which explains why low-density earlywood erodes more rapidly than denser latewood. There is a similar relationship between the different density of wood species and their rate of erosion during photodegradation. The penetration depth of light into wood depends on the wavelength of light as well. The wavelength dependence of light penetration was tested by the intensity of the IR band at 1730 cm^{-1} (Kataoka et al. 2007). Significant changes in the carbonyl band at 1730 cm^{-1} were detected at depths of up to 70, 100, 150, 250, 300 and 500 μm, respectively, for specimens irradiated with light of wavelengths of 278, 310, 341, 372, 403 and 434 nm, respectively. Light above 434 nm did not affect the intensity of this carbonyl band.

The most sensitive detector of the photodegradation is the colour change. The colour of wood is determined by the conjugated double-bond systems. These types of chromophoric groups are located in lignin and mostly in extractives. The colour change indicates the degradation of conjugated double-bond systems and the creation of such systems.

For investigating the colour change of wood, strong UV light emitter, mercury vapour lamp was used to irradiate specimens. The total electric power of the applied double mercury lamps was 800 Watt, and the samples were located 64 cm from the lamps. The emission spectrum of mercury lamp (supplied by the producer of the lamp) is located mainly in the UV region (80 %). It contains 31 % UV-A (380–315 nm), 24 % UV-B (315–280 nm) and 25 % UV-C (280–240 nm) radiation. The photons of UV-C range have sufficient energy to split almost all chemical bonds in wood. The temperature was constant 30 °C in the irradiation chamber avoiding thermal degradations. A previous study demonstrated the importance of the temperature during the artificial photodegradation (Persze and Tolvaj 2012). Hardwood and sapwood species were chosen with different extractive types and contents. The investigated hardwood samples were as follows: Ash (*Fraximus excelsior* L.), Black locust (*Robinia pseudoacacia* L.), Oak (*Quercus petraea*) and Poplar (*Populus x euramericana Pannonia*); the softwood samples were as follows: Larch (*Larix decidua* L.), Scots pine (*Pinus sylvestris* L.) and Spruce (*Picea abies* Karst.). Radial surfaces were prepared to determine the average changes in colour of earlywood and latewood. The colour of the wood specimens was measured before and after irradiations. The exposures were interrupted after 8, 20, 40, 90 and 200 h to measure the colour data.

The colour change was notable and visible even with the naked eye after two hours of irradiation for Black locust. In contrast, Poplar and Larch hardly suffered

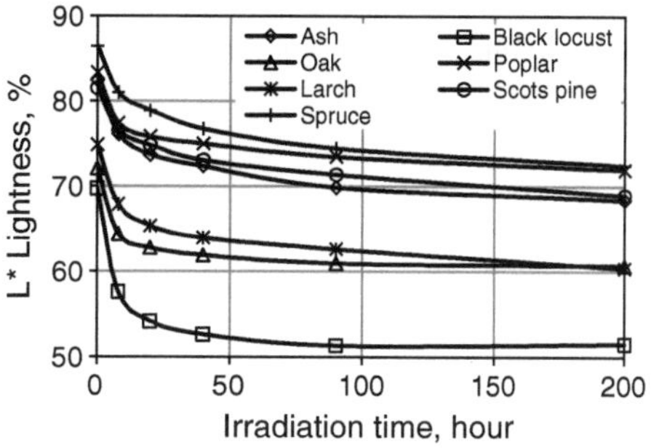

Fig. 2.32 The lightness change of the samples due to mercury lamp irradiation

any visible colour change during the same period of time. The lightness decreased rapidly during the first 10 h of irradiation (Fig. 2.32). After 20 h, the lightness change was moderate and almost linear. The Black locust suffered the greatest (18 units) and Oak and Poplar the smallest (11 units) lightness change during the 200 h of irradiation. The rapid change of lightness in the first 8 h (4 % of the total irradiation time) accounted for 67 % of the total change in Black locust and in Oak, 52 % in Poplar, 48 % in Larch, 45 % in Ash and 39 % in Scots pine and in Spruce. The lightness of Black locust and Oak did not change during the last 110 h of irradiation. The other samples produced slight, but continuous lightness decrease during this period.

The redness change (the alteration of a^* coordinate) showed much greater diversities among the species than the lightness change (Fig. 2.33). The redness increased continuously during the irradiation in all cases. The increase was slow and almost linear for all conifers. The Black locust and the Oak samples showed changes with great differences in intensity. The hardwood samples undergo a rapid increase during the first 8 h, and it was 67 % of the total change for Black locust, 33 % for Oak, 26 % for Ash and 20 % for Poplar. These data for conifers were between 6 and 13 %. Among the hardwoods, Poplar has less extractive content and it produced the smallest redness increase. The initial redness of Larch is higher than the same parameter for the other species and Larch generated the smallest change (3.7 units) during the 200-h treatment. It is thought that the dark coloured extractives are responsible for the high durability of Larch wood.

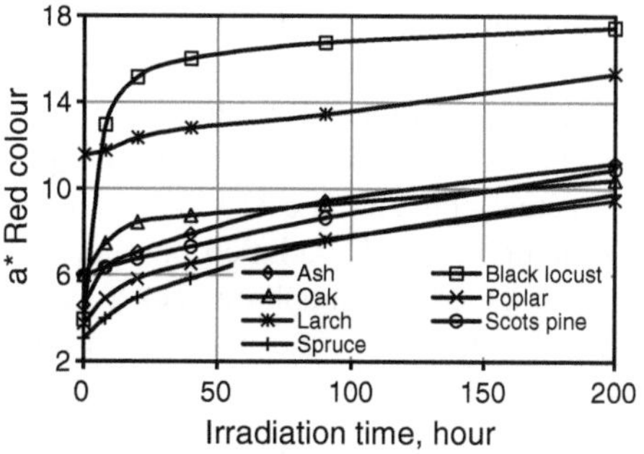

Fig. 2.33 The red colour change of the samples due to mercury lamp irradiation

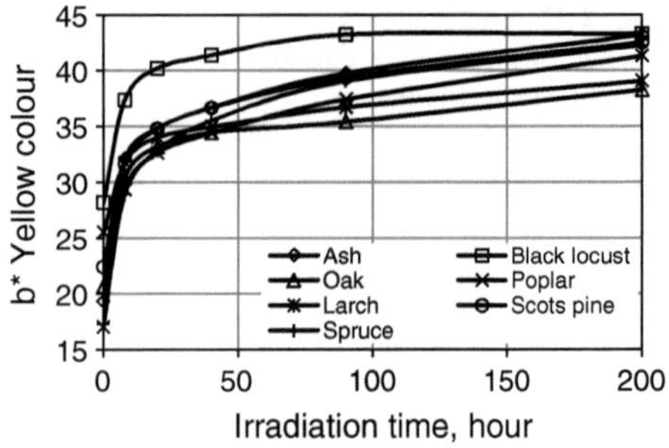

Fig. 2.34 The yellow colour change of the samples due to mercury lamp irradiation

The chromophoric groups in wood are located in the lignin, in the extractives and in their derivatives. Lignin derivatives play the main role mostly in yellowing. This phenomenon was widely investigated in terms of the yellowing of paper containing lignin. The degradation products of extractives mainly determine the alteration of red colour coordinate during photodegradation (Persze and Tolvaj 2012). Similar result was found related to the colour alteration by steaming as well. The Black locust wood has much higher extractive content than the other investigated species, and it contains robinetin-type extractives as well. The degradation products of robinetin produce an intensive increase in redness during steaming of Black locust (Tolvaj et al. 2010). That is why Black locust produced much greater increases of redness during photodegradation than the other species.

The yellow colour coordinate showed similar changes for all examined species (Fig. 2.34). The yellowness increased rapidly during the first 10 h of irradiation. After 20 h, the yellowing was moderate and the alteration was close to linear. The Black locust was an exception again. Its initial colour is more yellow (28 units) than the colour of the other species. Beside the high yellowness, it has almost no redness. That is why the natural appearance of Black locust wood is not attractive for most people. The initial yellowness of Larch (25.5 units) is also high, but it is accompanied by high red coordinate creating a pleasant warm brownish colour. The yellow colour coordinate of Black locust remained constant during the last 110 h of irradiation. The increase of yellow coordinate of Oak and Larch was less intensive than that of the other species during the last 180 h of irradiation (except Black locust). Poplar and Spruce suffered the greatest yellow colour change (24.5 units) during the 200 h of irradiation.

The uniform yellowing can be explained mainly with the photodegradation of lignin. The yellowing differences were generated by the protecting effect of the extractives and their derivatives.

The extraordinary behaviour of Black locust can be explained by its high extractive content. The UV light degrades the extractives followed by the rapid oxidation of the degradation products. This results in a rapid increase in redness. The modified chromophores act as a kind of energy trap which slow down the photodegradation of the main wood components, mainly lignin. Probably similar protection is provided by the originally existing chromophoric units of Larch wood.

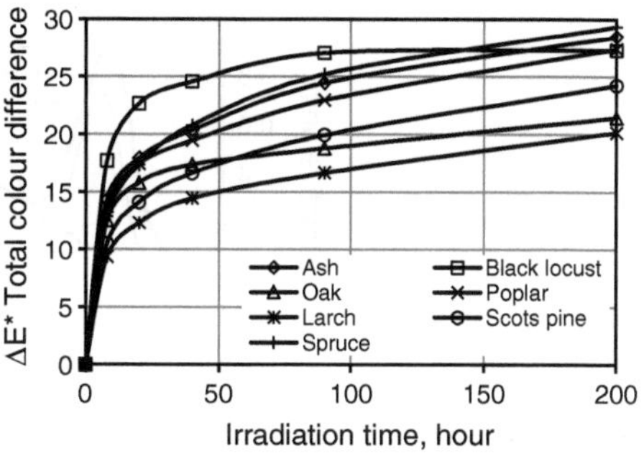

Fig. 2.35 The total colour change of the samples due to mercury lamp irradiation

Poplar and Spruce have low concentration of extractives. That is why these species produced the greatest yellow shift.

The total colour change as a complex parameter shows the summed colour alteration (Fig. 2.35). The trend lines show rapid colour alteration at the beginning of the treatment followed by moderate but continuous increase. The Black locust suffered greater colour change than the other species at the beginning of the treatment. During the second half of the treatment, Black locust did not show any colour change. The colour alteration of Oak was between Black locust and the other species during the second half of irradiation. The total colour difference data do not give additional information compared to the individual colour coordinates.

The alterations of the colour coordinates are sensitive indicators of the photodegradation but do not give exact information about the chemical changes. More detailed information can be obtained if the IR absorption spectrum is measured before and after treatment. The different chemical groups have individual absorption bands. An absorption increase indicates the increasing number of absorbing groups, while the absorption decrease represents the splitting of the chemical bounds. Wood is not transparent to light, so the direct measurement of absorption is not possible. During reflection, the light penetrates slightly into the wood surface with a little absorption and scattering. The diffusely reflected light provides information about the absorption.

Using this method, the diffusely reflected light is collected and its absorption is calculated by the K-M equation (Eq. 2.32). Fortunately, the scattering coefficient hardly depends on the wavenumber of the IR light; thus, the shape of K-M function is similar to the wavenumber dependence of the absorption coefficient. This K-M function is usually used as the IR absorption spectrum.

The infrared spectroscopy is an excellent tool to determine the chemical changes caused by the light irradiation. It provides information from the surface layer affected by the irradiation. The fingerprint region contains the highly overlapped absorption bands of cellulose, hemicelluloses and lignin. Figure 2.36 represents the absorption spectra (calculated by the K-M theory) of Beech sample before and after 24-h irradiation by mercury lamp at 70 °C. The carbonyl band between 1680 and 1820 cm^{-1} increased, while the peaks of the aromatic ring vibrations arising from lignin (at 1508, 1460 and 1425 cm^{-1}) decreased during the irradiation. The band for

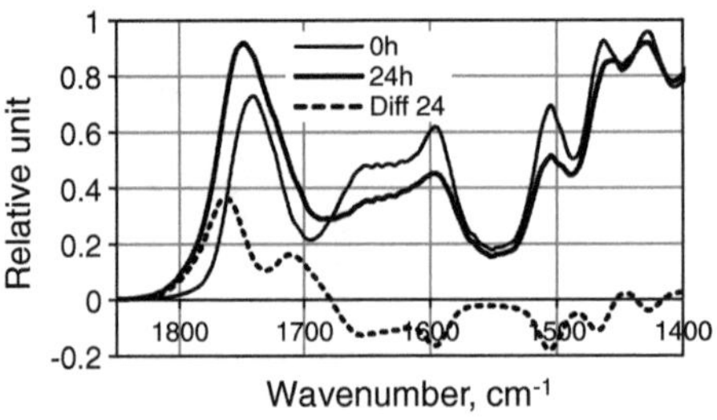

Fig. 2.36 Absorption spectra of Beech before treatment and after 24-h UV treatment and the difference spectrum

the aromatic skeletal vibration at 1596 cm^{-1} also decreased, but it can be observed in the case of hardwoods only. This peak does not exist in the case of softwoods because it belongs to the so-called syringyl lignin found mostly in hardwood. Bands at 1596, 1504, 1455 and 1430 cm^{-1} have similar width, but the carbonyl band at 1741 cm^{-1} is much wider than the others. It suggests that this carbonyl band is the superposition of some individual bands. Hence, most publications treat the band of unconjugated carbonyls as a single absorption band with a maximum around 1740 cm^{-1}.

If we have a closer look at the absorption change of the unconjugated carbonyl band (Fig. 2.36), the shift of the band towards higher wavenumbers is visible. The maximum moves to 1748 cm^{-1}. Moreover, the greatest difference between treated and initial spectra appears not at the maximum but at the left side of the band. The difference at the maximum (1740 cm^{-1}) is 0.14, and at 1763 cm^{-1}, it is 0.37 units, which is more than double. These findings highlight the disadvantage of the simple comparison method where the initial and the treated spectra are presented on top of each other.

The difference spectrum method solves this problem. Creating the difference spectrum (irradiated minus initial), only those absorption bands appear where change has happened. The absorption increase is represented by positive band, while negative band represents the absorption decrease. The changes are clearly visible by calculating the difference spectrum of Beech sample irradiated for 24 h (Fig. 2.36 dotted line).

The difference spectrum of Beech consists of two positive bands at 1765 and at 1705 cm^{-1}. This result is strengthened by 2D IR spectroscopy (Calienno et al. 2014). The band at 1763 cm^{-1} represents the absorption of CO stretching for unconjugated ketones and γ lactones generated by the oxidation after the splitting of the aromatic ring. The band at 1705 cm^{-1} represents the absorption of aliphatic carboxyl groups (Tolvaj et al. 2013). Between 1600 and 1690 cm^{-1}, the conjugated carbonyls in lignin and in phenolic molecules and quinone carbonyl groups are absorbed, and the absorption of OH groups in water is also present. In this region, Beech shows a broad decrease with two minimums at 1655 and at 1596 cm^{-1}.The absorption decrease at 1596 cm^{-1} indicates the degradation of aromatic ring in

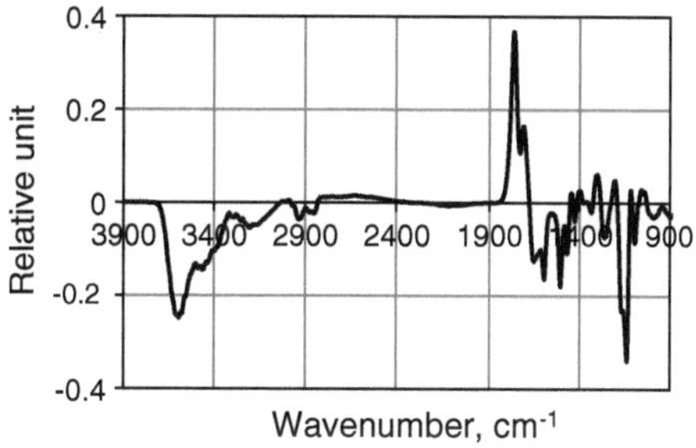

Fig. 2.37 Difference spectrum of Beech caused by 24-h UV treatment

syringyl lignin. The negative peak at 1508 cm^{-1} belongs to the aromatic skeletal vibration of guaiacyl lignin. This negative peak is detectable together with the absorption decrease of the aromatic C-H deformation at 1469 and 1428 cm^{-1} and with the absorption decrease of the guaiacyl ring breathing at 1270 cm^{-1} (Fig. 2.37). The degradation of the aromatic ring for lignin is demonstrated by these absorption decreases. The absorption changes of Beech wood are presented in Fig. 2.37 within the whole examined region. The greatest absorption decrease is visible at 1174 and 1139 cm^{-1}. The first decrease belongs to the asymmetric stretching of ether bond in cellulose. The second decrease belongs to the symmetric stretching of ether bond, the aromatic C-H deformation and to the glucose ring vibration. These absorption decreases indicate the ether splitting and the depolymerisation of cellulose (Tolvaj et al. 2013).

There are 4 small positive bands in the 1100–1450 cm^{-1} wavenumber range. The accuracy of these peaks (at 1445, 1309, 1211 and 1119 cm^{-1}) is questionable. They may be created by the K-M transformation (Tolvaj and Mitsui 2011).

There are absorption decreases in the OH absorption region (3000–3700 cm^{-1}). This is a wide band as a result of the superposition of many individual absorption changes belonging to the OH groups located in different positions. The minimum of the absorption decrease is at 3600 cm^{-1}. The broad band at 3600 cm^{-1} presents the absorption decrease of the intramolecular hydrogen bond in the phenolic group in lignin. This absorption decrease can be partly the effect of drying, i.e. the elimination of the water from the wood. This intensity loss can also be interpreted as the rupture of intramolecular OH bonds of cellulose, comparable to that of the ether bridge. The absorption decrease in OH region can also be interpreted by the loss of OH groups followed by dehydration after UV irradiation. Dehydration belongs to the plausible reaction, during which double bonds—responsible for coloured compounds—are formed. The formation of esters and lactone rings, also common features of photodegradation, occurs via water loss and OH group consumption (Tolvaj and Faix 1995).

There are two small absorption decreases at 2846 and 2936 cm^{-1} belonging to the symmetric and asymmetric methoxyl stretching. These small decreases are thought to be triggered by the demethylation, elimination of CH_2OH groups both from lignin and polysaccharides.

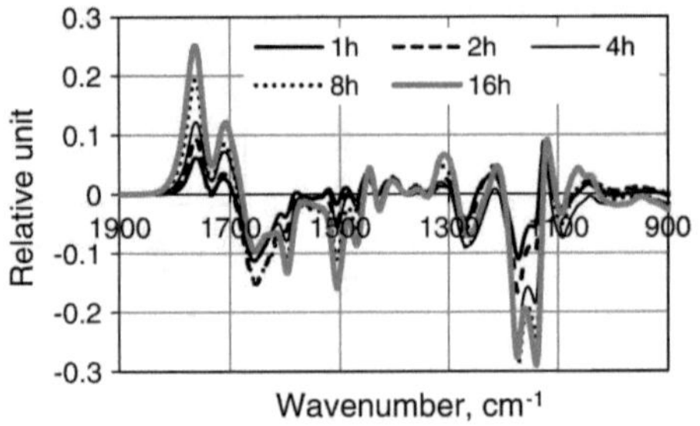

Fig. 2.38 Difference spectra of Beech after 1, 2, 4, 8 and 16 h of UV irradiation by mercury lamp

The time dependence of photodegradation can be monitored by the difference spectrum method. Figure 2.38 represents the difference spectra of Beech after 1, 2, 4, 8 and 16 h of UV irradiation by mercury lamp. The lignin degradation was continuously demonstrated by the absorption decrease at 1506 and 1596 cm^{-1}. The negative peak at 1654 cm^{-1} first decreased than remained constant. The two carbonyl peaks around 1707 and 1763 cm^{-1} increased parallel. The growing of the peak at 1763 cm^{-1} was a little faster than the growing of the nearby peak. The most remarkable change happened at 1174 and 1139 cm^{-1}, respectively. The decrease of the peak at 1174 cm^{-1} was faster at the beginning, but it slowed down after 4 h. The other peak at 1139 cm^{-1} decreased continuously and has overtaken its neighbour before 16 h of irradiation. The difference between the two peaks increased forth after 24 h of irradiation (see also Fig. 2.37).

There are differences in the course of photodegradation depending on the wood species. Figure 2.39 represents the IR absorption changes of Beech, Spruce and Oak earlywood caused by irradiation for 8 h at 30 °C by mercury lamp. The absorption decrease of guaiacyl lignin at 1508 cm^{-1} is similar for all species. However, the absorption decrease of syringyl lignin at 1596 cm^{-1} is missing in case of conifers. The reason is that conifers have hardly any syringyl lignin. The absorption increase in the region of carbonyl groups (1670–1810 cm^{-1}) is mostly originates from the degradation products of lignin. Therefore, there are differences between conifers

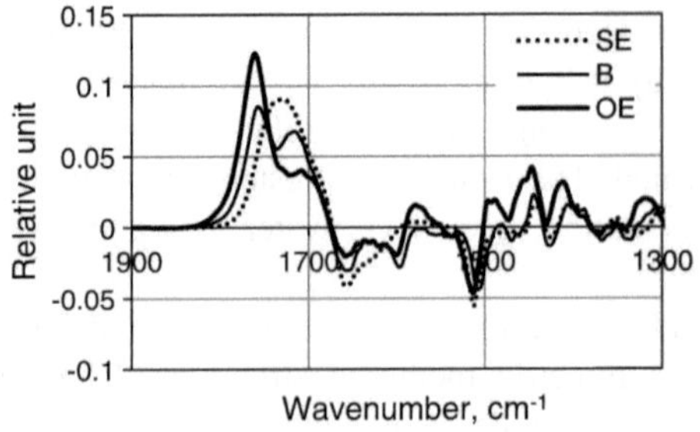

Fig. 2.39 Difference spectra of Beech (*B*), Spruce (*S*), Oak (*O*) and earlywood (*E*) after 8-h UV irradiation

and deciduous species in the carbonyl region as well. Beech has two separate maxima at 1710 and 1760 cm^{-1}. Oak earlywood has also two maxima similar to Beech, but the intensities are different. The peak at 1760 cm^{-1} is much higher than the other. Probably, Spruce earlywood has two positive bands, but only one at 1710 cm^{-1} is the same for all three species. The other peak moved closer, creating a visually integrated band at 1730 cm^{-1} for Spruce. The situation is the same for all conifers. The width of this band confirms that it is the superposition of two absorption increase. These results suggest that the absorption increase at 1710 cm^{-1} is created mainly by the degradation products of guaiacyl lignin and the other absorption increase at 1760 cm^{-1} is created mainly by the degradation products of syringyl lignin. There must be a third peak arising between the two previous ones. These suggestions, however, need further chemical investigations.

There are differences in sensibilities of earlywood and latewood to the photodegradation. The latewood has usually more protecting ability than the earlywood. It is well demonstrated in Figs. 2.40 and 2.41. Figure 2.40 demonstrates the absorption change difference between Poplar earlywood and latewood. The lignin in earlywood is more degradable than that in the latewood. The absorption decrease of both syringyl (at 1595 cm^{-1}) and guaiacyl (at 1505 cm^{-1}) lignin is smaller in latewood than that in earlywood. As a consequence, the absorption increase of carbonyl groups is smaller in latewood than that in earlywood. The difference is greater at the peak of 1706 cm^{-1} than that at 1760 cm^{-1}. There is one place where the latewood suffered greater changes than the earlywood, namely in the absorption band of the asymmetric stretching of ether bridge at 1172 cm^{-1}.

Conifers have somewhat different photodegradation behaviour than deciduous species (see Fig. 2.39). Figure 2.41 shows the changes in absorption differences of Spruce earlywood and latewood. The conjugated carbonyls (around 1660 cm^{-1}) in latewood suffered greater degradation than the same groups in earlywood. This broad negative peak pulled down the partly overlapped positive peak at 1706 cm^{-1}. There is a small negative baseline shift for latewood in this region as well. The earlywood was less stable than the latewood indicated by the absorption decrease at 1507 cm^{-1}. The small positive peak at 1490 cm^{-1} pushed the negative peak towards

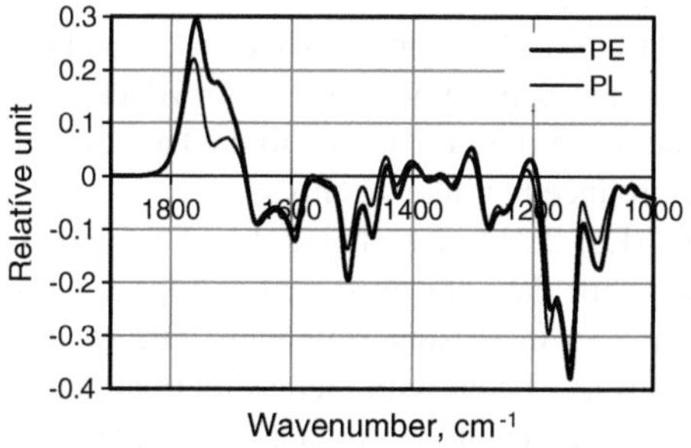

Fig. 2.40 Absorption change of earlywood (*E*) and latewood (*L*) for Poplar (*P*) caused by 16-h UV treatment

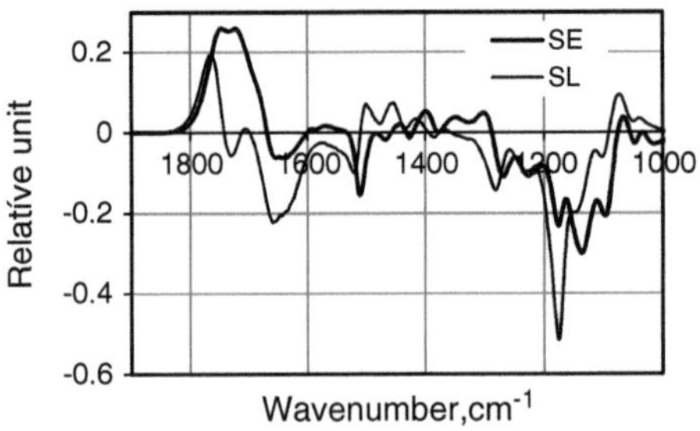

Fig. 2.41 Absorption change of earlywood (*E*) and latewood (*L*) for Spruce (*S*) caused by 24-h UV irradiation

higher wavenumbers for latewood. The latewood produced smaller absorption increase in the non-conjugated carbonyl region than the earlywood. Both peak intensities (at 1713 and 1736 cm^{-1}) are the same for earlywood. There is a hardly visible shoulder at 1680 cm^{-1} representing the third absorption increase caused by UV irradiation. However, for latewood, the peak at shorter wavenumber is much smaller than the other at greater wavenumber. This behaviour of spruce is similar to Poplar (see Fig. 2.40). The peak where the latewood suffered greater change than earlywood is located at 1175 cm^{-1}. The situation is similar for Poplar as well.

Examining wood species having heartwood and sapwood, it was realised that the heartwood production does not influence the photodegradation behaviour of earlywood and latewood. There was no remarkable difference in the photodegradation of earlywood sapwood and heartwood. The same was found for latewood.

The photodegradation increases the surface roughness of wood. The roughness was measured with two methods: optical and contact surface roughness measurement. The unfiltered P profile was computed from the traced profile (EN ISO 3274). The surface roughness parameters (Pz; Pt; Pa; Pmax) can be used also for this profile. The roughness parameters were recorded after five days of UV light irradiation, which was followed by two-day intervals. There was a total of 6 irradiation periods. The necessary parameters were measured right after subjecting the samples to light irradiation. The roughness parameters increased continuously during UV irradiation for all examined species. Figure 2.42 shows the results for Poplar measured on radial surfaces. The roughness increase was linear, and similar results were found for Spruce and Scots pine as well. For Black locust, the increase was a little higher during the first 5 days compared to the later period (Tolvaj et al. 2014a). The optical method produced similar results.

The baseline shift (Eq. 2.31) of earlywood and latewood for Poplar was measured. These baseline shift data are presented in Fig. 2.43. The optical roughness measurement also showed increasing roughness due to the UV irradiation. The increase was continuous and linear with increasing treatment time. This change was in good correlation with the results of contact roughness measurement. Figure 2.43 represents the high sensitivity of the optical roughness measurement. The latewood became less rough than the earlywood. This figure shows clearly that earlywood is

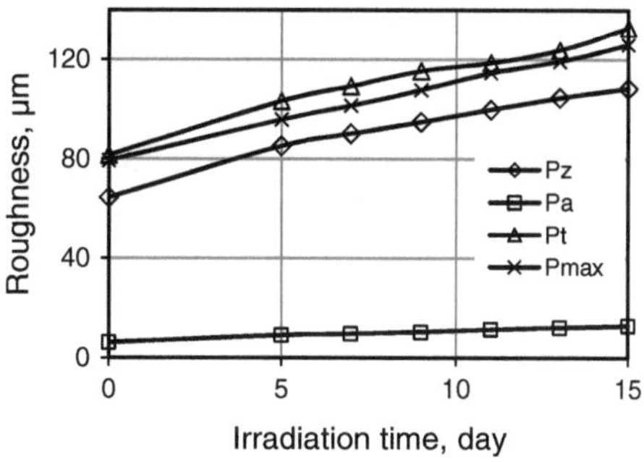

Fig. 2.42 Selected roughness parameters of UV-treated Poplar wood

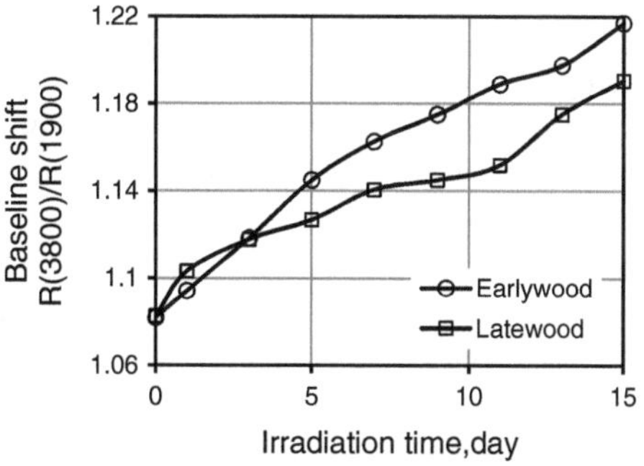

Fig. 2.43 IR baseline shift of Poplar caused by UV irradiation

more sensitive to the UV irradiation than the latewood. This finding is in good correspondence to the results of IR changes presented in Fig. 2.40. The IR absorption measurement also indicated that earlywood suffered greater chemical changes during UV irradiation than the latewood. Figure 2.43 shows that the applied optical roughness measurement is highly sensitive to represent the roughening effect of photodegradation.

2.5.3 Photodegradation of Thermally Modified Wood

There are two main types of thermal treatment. One is the treatment at low temperature (80–130 °C) under wet condition. It is called steaming, and the steam must be saturated. If the steam is not saturated (overheated steam), the wood will dry rapidly. The other possibility is the dry thermal treatment where the temperature must be higher than for steaming. The dry thermal treatment causes considerable changes if the temperature is above 150 °C. Below this temperature, the chemical changes are slow according to the Arrhenius law (2.33). Thus, it would take a long time to get the proper dark colour under dry condition. The cost is a disadvantage of low temperature for industrial use. However, chemical changes become faster with rising temperature. This increase is exponential according to the Arrhenius law.

The chemical decomposition of wood is highly intensive above 220 °C, and it causes rapid decrease of the excellent mechanical properties of wood. Therefore, industrial thermal treatments apply temperatures below 220 °C. The wood material treated under dry thermal condition is usually called thermowood. The colour of a material is created by the conjugated double-bond (chromophore) chemical systems. Lignin and certain extractives consist of such chromophores in the natural wood. Thermowood contains additional chromophores produced during the degradation of the hemicelluloses.

To determine the short-term photodegradation behaviour of thermowood samples, two hardwood and two softwood species were utilised. The hardwood species were Black locust heartwood (*Robinia pseudoacacia* L.) and Poplar (*Populus x euramericana Pannonia*). The softwood species were Larch heartwood (*Larix decidua* L.) and Spruce (*Picea abies* Karst.). The initial moisture content of the samples was between 8 and 10 %. The dry thermal treatment was carried out in a laboratory scale drying chamber at 160 and 200 °C in the presence of air and ventilation (open system). The preheating time was 30 min. The duration of the effective thermal treatment time was 2, 4 and 6 h. The 2-h treatment time at 160 °C produced only a minor colour change; therefore, these samples were not used for further investigation. The thermally treated samples along with untreated control samples were then artificially photodegraded. A strong UV light emitter, mercury vapour lamp provided the light irradiation. The L^*, a^* and b^* colour coordinates were measured and calculated based on the D_{65} illuminant and 10° standard observer with a test-window diameter of 8 mm (Tolvaj et al. 2014b).

The samples underwent considerable colour change during thermal treatments, as shown in Figs. 2.44, 2.45, 2.46, 2.47 and 2.48. These data are presented in the figures at zero irradiation time. The lightness of all species decreased compared to the untreated samples. The magnitude of change followed the intensity of the thermal treatment. In other words, higher temperature and longer treatment time produced greater decrease in lightness. The selected thermal treatment parameters resulted in a large variety of lightness (see the first dots in Fig. 2.48). The redness of all species increased with heating, and the order of the change followed the intensity of the thermal treatment (see Fig. 2.44). Higher temperature and longer treatment time produced a greater increase in redness. Spruce was highly sensitive to both thermal treatment temperature and duration of exposure. Poplar was also highly sensitive to the temperature of the thermal treatment, but not sensitive to the thermal treatment time. The yellow colour of Larch, Poplar and Spruce increased following the tendency of red colour change (Fig. 2.46). The only exception was the Black locust (Fig. 2.45). Its originally high yellow colour decreased due to the thermal treatment following the same tendency as the lightness change.

The photodegradation due to UV irradiation caused the most detailed differences in the redness change. The red colour changes are represented in Figs. 2.44. Trend lines showing the red coordinate shift were convergent for the low extractive content species (Poplar and Spruce), but they were first rapidly convergent and later divergent for the high extractive content species (Black locust and Larch). Black locust and Larch samples produced rapid change in redness during the first 3 h of

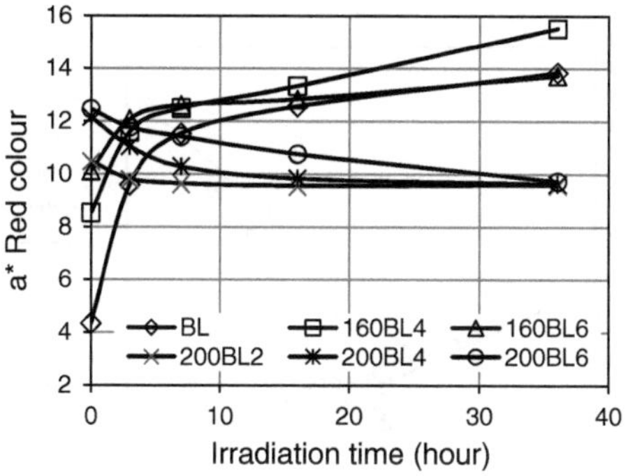

Fig. 2.44 Redness change of Black locust caused by light irradiation (*abbreviation* thermal treatment temperature/sample name/thermal treatment time in hour)

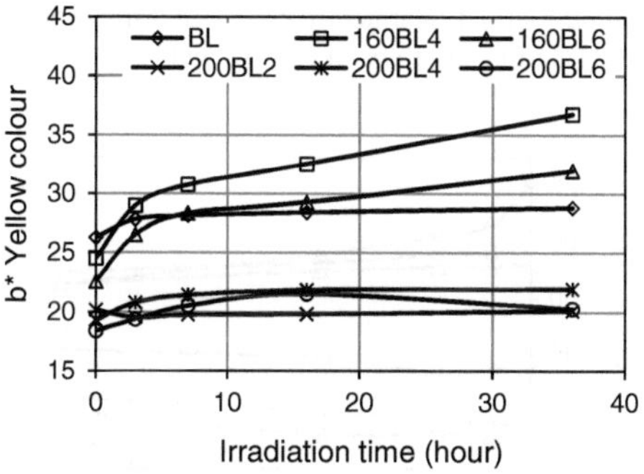

Fig. 2.45 Yellow coordinate shifts of Black locust samples caused by light irradiation (*abbreviation* thermal treatment temperature/sample name/thermal treatment time in hour)

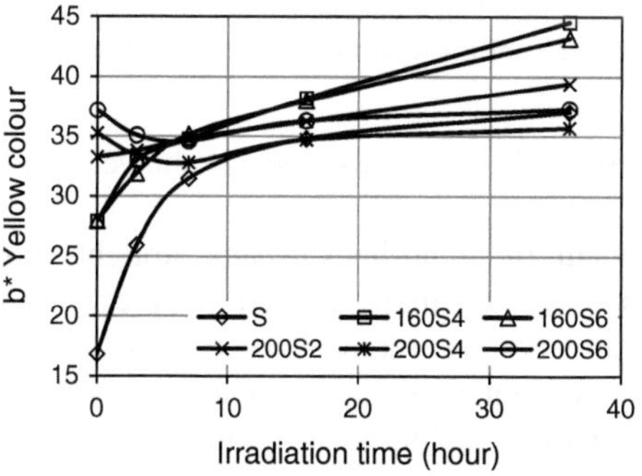

Fig. 2.46 Yellow coordinate shifts of Spruce samples caused by light irradiation (*abbreviation* thermal treatment temperature/sample name/thermal treatment time in hour)

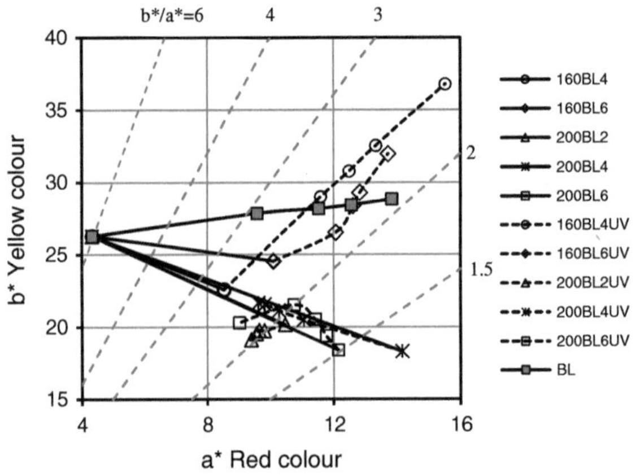

Fig. 2.47 Redness and yellowness change of Black locust (BL) during dry thermal treatment and UV irradiation (*abbreviation* thermal treatment temperature/sample name/thermal treatment time in hour). The duration of UV irradiation was 6, 16, 36 and 72 h

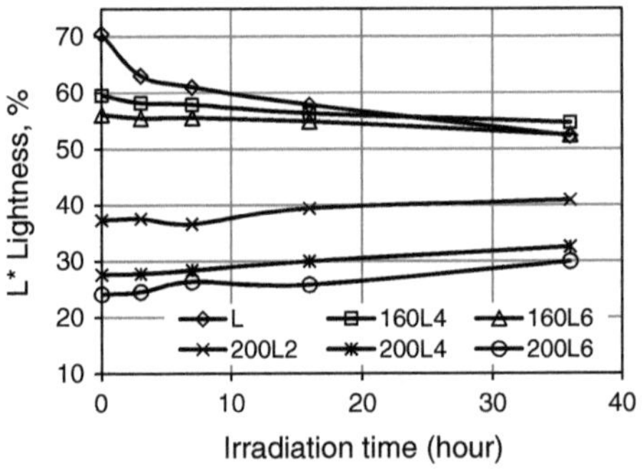

Fig. 2.48 Lightness change of the Larch samples caused by light irradiation (*abbreviation* thermal treatment temperature/sample name/thermal treatment time in hour)

exposure. In comparison, the other two species showed moderate redness change during this period.

After 7 h of treatment, the trend lines became straight, showing a slow increase or remained constant. The thermally treated samples showed different changes depending on the treatment temperature. Red coordinate shift of samples treated at 160 °C followed the tendency observed in natural wood (thermally untreated) samples. This allows us to conclude that the thermal treatment at 160 °C did not affect the redness change behaviour of wood.

However, the samples thermally treated at 200 °C changed their redness in a different way than untreated wood. After moderate decrease in redness during the first 7 h of light irradiation, these treated samples kept a nearly constant redness value during the examined period. The samples treated at 200 °C for 2 h showed the most stable redness values during photodegradation. The redness change is mainly determined by the chemical changes of extractives and by the chromophore products of the thermally degraded hemicelluloses. It can be concluded that at 200 ° C thermally modified extractives and the chromophore products of hemicelluloses are fairly stable during light irradiation.

The yellow colour changes of the examined species are presented in Figs. 2.45 and 2.46. Two graphs are presented here because the changing tendencies and values of Poplar were the same as those of Spruce and similar to Larch. Larch samples showed similar behaviour as Spruce samples; only the intensity of the changes was a little stronger for Larch during the first 3 h of light exposure. The yellow colour of Black locust is naturally different compared to the other European species. This deviation is caused by its high robinetin content. A previous study demonstrated that the high extractive content of Black locust partly protects its lignin content during light irradiation (Tolvaj and Varga 2012).

The same finding is partly demonstrated here as well. The untreated Black locust produced the least yellowing compared to the other examined untreated samples. Some of the extractives of Black locust underwent thermal degradation, represented by the increasing values of a^* and b^* coordinates. The extractives modified during the thermal treatment at 160 °C were not able to protect the lignin of Black locust, as it is visible on Fig. 2.45. The yellow colour increase of the Black locust samples treated at 160 °C was much greater than that of the untreated samples. The degradation behaviour of thermally modified Black locust at 160 °C was similar to the degradation of the other species that do not contain protective extractives. Black locust samples treated at 200 °C did not produce remarkable yellowness changes. The other samples showed a slight yellow colour decline during the first 7 h of light irradiation. After this time period, the yellowing of the samples treated at 200 °C followed the slow yellow coordinate increase of the untreated samples. The yellow colour change of the samples treated at 200 °C for 2 h seemed to be the most stable, as discussed in the case of red colour change. The yellow colour change of the investigated samples showed that the lignin of thermally modified wood underwent a similar photodegradation as that of the untreated natural wood. This was found in a previous study (Srinivas and Pandey 2012).

There is another possibility to represent the colour alteration on the a^*- b^* plane (Fig. 2.47). The equal b^*/a^* lines (constant hue) are also represented in this figure. The thermal treatment (solid line) generated moderate decrease in yellowness and great increase in redness for Black locust. The UV treatment generated considerable colour saturation for the samples pretreated at 160 °C. The samples pretreated at 200 °C did not change the saturation, but the hue returned to those samples which were pretreated at 160 °C. The colour hue was in a narrow interval between 2 and 2.5 during the UV treatment. The thermal treatment at 200 °C pushed the value of hue below 2, but the UV irradiation raised back this value above 2. The effect of the two type of pretreatments was opposite for the effect of UV treatment. The other wood species produced similar colour changes during the UV irradiation; only the redness change was much less and the yellowness change was positive during the thermal pretreatment.

The lightness of the investigated samples decreased considerably during the dry thermal treatment (Fig. 2.48). This decrease was 83, 66, 62, and 56 % for Black locust, Larch, Poplar and Spruce, respectively.

There were only minor differences among the species during the light irradiation. That is why only the data of Larch are presented here. Poplar and Spruce showed the same changing behaviours, and this pattern of change was also similar to that of the Larch samples. Untreated wood samples suffered rapid lightness decrease during the first 10 h of light irradiation, followed by moderate decrease. The originally dark thermally treated samples exhibited slow but continuous changes. The values of these changes were considerably less than those of the untreated samples. The direction of change was determined by their initial lightness value. Samples having initial lightness values higher than 40 units showed decreasing lightness, while those with initial L^* values less than 40 units showed lightening during irradiation. These results suggest that thermal treatment slightly reduced the lightness change effect of photodegradation (Tolvaj et al. 2014b).

The short-term photodegradation behaviour of thermowood is valuable for the scientist, but in practice, the long-term effects are even more important. Some researcher found that dry thermal treatment increases the weathering stability of wood, while others stated that thermowood undergoes the same photodegradation as natural wood. Some results are presented here.

Ayadi et al. (2003) tested the colour stability of heat-treated wood samples. The heat treatment was performed at 240 °C for 2 h in nitrogen atmosphere. Heat-treated samples of Ash, Maritime pine and Poplar heartwood were exposed to UV light for 835 h. The total colour change was determined as a complex parameter representing the changes. The experiments showed that the colour stability of the heat-treated wood was better after the 835 h of exposure when compared to untreated wood. The difference in degradation between thermally treated and untreated wood increased rapidly during the first 36 h of UV light exposure. After this period, the trend lines of total colour change were close to parallel.

Huang et al. (2012) investigated the photodegradation of heat-treated (190–115 °C) Jack pine (*P. banksiana*), Aspen (*P. tremuloides*) and Birch (*B. papyrifera*) samples. There was a major increase in yellow and red colour coordinates with corresponding decreases in the reflectance (L^*) for all the three species during heat treatment. Heat-treated woods had better colour stability during the early times of weathering, while the colours of heat-treated woods and untreated woods were very similar after the specimens have been subjected to long-term artificial weathering. They found that degradation of hemicelluloses can consequently cause an increase in the amount of lignin component of Birch wood. This increase becomes more important as the treatment temperature increases. It was also found that the heat treatment increases the crystallised cellulose content. The increased lignin and crystallised cellulose content brought about by heat treatment protects heat-treated birch against degradation taking place during weathering (Huang et al. 2013).

Yildiz et al. (2011) investigated the outdoor weathering properties of heat-treated Alder wood. The treatment parameters were as follows: 150, 180 and 200 °C, for periods of 2, 6 and 10 h, respectively. The samples were outside for 3 years in Turkey. It was found that heat treatments delayed and decreased the rate of colour change caused by the weathering but did not completely prevent it. The most advantageous treatment condition was at 200 °C for 10 h. The photodegradation behaviour of thermally treated Rubber wood (*Hevea brasiliensis*) was investigated by Srinivas and Pandey (2012). Oven-dried samples were kept in a glass tray and were heat treated in a preheated vacuum oven at 225 °C for 2, 4 and 6 h under 400 mm (Hg) of vacuum to obtain three different levels of thermally modified wood specimens. Results of colour changes and FTIR spectroscopy revealed that thermal modification of wood does not induce resistance against UV radiation.

Pannonia poplar (*Populus x euramericana Pannónia*) samples were thermally treated in vegetable oil at 160 and 200 °C (Bak et al. 2012, 2014). After thermal treatment, the samples underwent outdoor weathering during a year. It was concluded that wood samples treated at lower (160 °C) temperature underwent more severe total colour changes compared to the samples thermally treated at higher temperature (200 °C). It means that the colour stability of the thermally treated wood depends mainly on the temperature.

2.6 Weathering Processes of Wood

Outdoor weathering processes of wood can be divided into two main groups: biotic and abiotic degradations. The biotic degradation is mainly caused by fungi and insects attack. Biotic degradations will not be discussed in this book.

The main factor of the abiotic degradation is the sunlight. Within the sunlight, the UV radiation is the most harmful for biomaterials. The results of UV radiation for wood were discussed above. The effectiveness of UV radiation is partly influenced by the temperature of the wood. This effect was neglected earlier, but nowadays some publications highlight its importance. In addition, rain leaches out the water soluble extractives and degradation products from the surface of the outdoor wood.

The colour of individual wood species is mainly defined by its extractive content. The degradation of these extractives partly determines the colour alteration during photodegradation. Pandey (2005b) compared the photodiscolouration of natural and extractive free wood samples. He demonstrated that the extractive free specimens exhibited a steady increase in colour change with increasing irradiation time. Unextracted wood surfaces showed rapid colour changes during the initial period of exposure, which decreased upon prolonged exposure time. Chang et al. (2010) found that the rate of wood degradation was lessened by the presence of extractives, when they examined the photodegradation of Japanese cedar and Taiwan acacia.

2.6.1 Photodegradation Due to Sun Radiation

The direct investigation of photodegradation caused by sunlight is difficult. Exposure to direct sunlight is not repeatable under the same conditions. Rainfall, air humidity and alteration of the temperature are uncontrollable parameters that are able to modify the effect of sunlight. The other problem is the changing intensity of the sun radiation during the day and over the year. Clouds sometimes partly or fully cover the sun. These periods should be subtracted from the total treatment time, but it would be extremely difficult. It means that the irradiation time and intensity (the most important parameters) are not exactly known during outdoor solar radiation. Therefore, the effect of light irradiation is mostly investigated using artificial light sources.

In our experiments, the samples were exposed to sunlight exposure only on sunny days to determine the effect of the sunlight alone (Tolvaj and Mitsui 2005). That type of outdoor irradiation was not found in the literature at all. The aim of the study was to investigate the pure photodegradation effect of sunlight. The other aim was to determine whether artificial light sources are able to properly imitate the photodegradation of wood caused by sunlight. The similarities and the differences in the effects caused by natural and artificial light (xenon and mercury lamps) on wood were examined, to determine which type of radiation simulates the effects of sunlight. The air temperature varied from 16 to 41 °C. The max. r.h. was 80 %, and the daily average of total solar power density was between 436 and 459 W/m^2. The SX-75 Suga Test Instrument was equipped with a xenon lamp. The power density of the lamp was 180 W/m^2 within the range of 300–400 nm, at 63 °C (black panel) and 50 % r.h. As the intensity of sunlight changes continuously during the day and during the year, the comparison was made on the basis of irradiation time, not on the irradiated energy. Total irradiation time was 200 h for both sunlight and xenon light.

During the first 30 h of light irradiation by sun and xenon lamp, the colour change was rapid, as it is shown in Figs. 2.49, 2.50, 2.51 and 2.52. Changes in lightness were particularly pronounced (Fig. 2.49). The first 30 h of sunlight accounted for ca. 60 % of the total change of lightness in hardwoods and 40–50 %

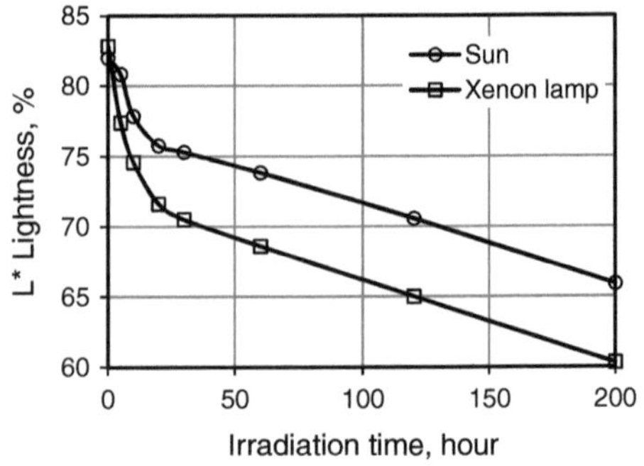

Fig. 2.49 Lightness change of Spruce due to irradiation by different light sources

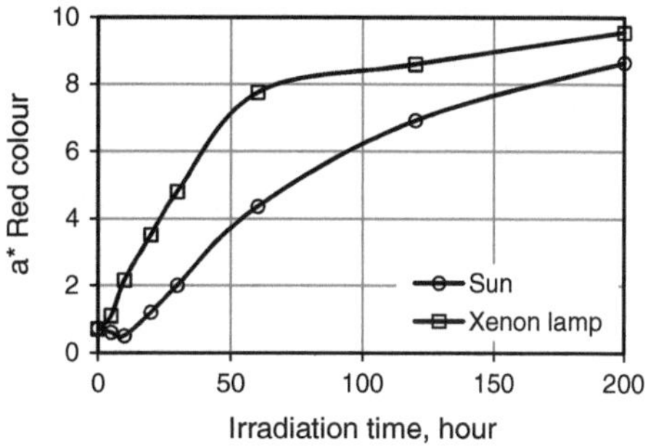

Fig. 2.50 Red colour change of Spruce due to irradiation by different light sources

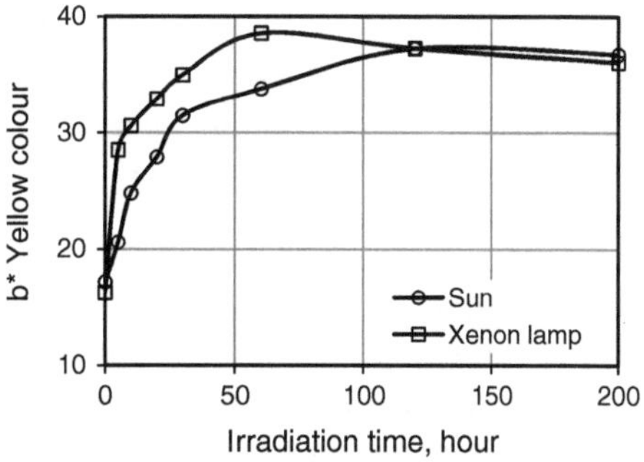

Fig. 2.51 Yellow colour change of Spruce due to irradiation by different light sources

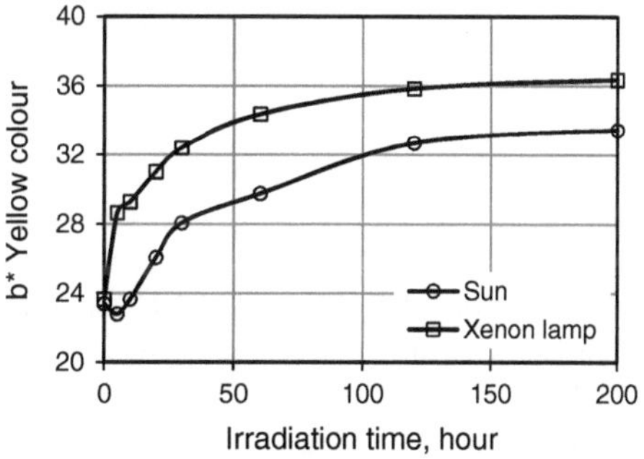

Fig. 2.52 Yellow colour change of Poplar due to irradiation by different light sources

of that in softwoods. Artificial light sources caused even faster changes in lightness than sunlight during the first 30 h of exposure. After 30 h of irradiation, the trend changed and the trend lines became parallel.

The a^* coordinate represented in Fig. 2.50 for Spruce did not change significantly during the first hours of exposure to sunlight. Even in the first 10 h of irradiation, sunlight produced only a minor redness decrease in Spruce just as in the other conifers. Only Black locust became redder intensively during the first hours of sunlight exposure. In contrast, samples exposed to xenon light became redder already in the first hours of irradiation. Black locust produced half of the total change in redness during the first 10 h. The extraordinary behaviour of Black locust may be the evidence of its high extractive content. The robinetin content of this species was especially sensitive to UV light exposure. The xenon lamp created greater increase in redness up to 60 h of irradiation than sunlight. After this period, the difference decreased continuously.

The third colour coordinate b^* for Spruce shows even faster alteration than a^* if sunlight and xenon light exposure are compared (Figs. 2.51 and 2.52). Yellowing of the samples caused by xenon light was concentrated in the first 5 h of irradiation. The yellowing during this period accounts for 61 % of the total yellowing of Spruce sample induced by xenon lamp. The same change caused by sunlight accounts only 17 %. After 60 h of irradiation, the intensity of change decreased or stopped and the trend lines converged or became parallel (Tolvaj and Mitsui 2005). Poplar showed less yellowing than the Spruce sample independently of the light source. The yellowing is produced mostly by the oxidation products of lignin degradation. The lignin decay and the resulting quinone formation are related to the photoyellowing (Müller et al. 2003). These newly created quinones are also sensitive to light irradiation and undergo further degradation. In the first period of irradiation, the quinone production is dominant and slow down later on. On the other hand, the degradation procedure of quinone structures is slow at the beginning (there are only a few) but will become more intensive later on. The result of these two processes produces a maximum. After reaching the maximum, the creation of quinone structures is slower than their degradation. For Spruce, this maximum is located at 60 and 120 h of exposure created by xenon lamp and sun irradiation, respectively. For Poplar, these maxima are located out of the examined time range.

The xenon lamp irradiation simulates the colour change caused by sunlight only when wood is subjected to a long-term exposure. Short-term exposure to xenon light causes a more rapid colour change, mainly yellowing, than sunlight.

Infrared spectroscopy was used for studying the chemical changes in wood caused by light irradiation. The differences in the change of IR spectra caused by the two types of irradiation (sun and xenon lamp) are qualitative rather than quantitative. The difference IR spectra were calculated to clearly present the changes. Deviation was found between hardwoods and softwoods represented here by the spectra of Spruce and Black locust in Figs. 2.53, 2.54 and 2.55, respectively. Only the fingerprint area is displayed because the main differences were found in this region. The band assignment can be found in Table 2.2.

After irradiation, the carbonyl band between 1680 and 1900 cm^{-1} increased. There was a typical negative peak at 1508 cm^{-1} belonging to the aromatic skeletal vibration of guaiacyl lignin. This negative peak is visible together with the absorption decrease of the aromatic C–H deformation at 1469 and 1428 cm^{-1} and with the absorption decrease of the guaiacyl ring breathing at 1265 cm^{-1}. The band for the aromatic skeletal vibration at 1600 cm^{-1} (syringil lignin) also decreased in the case of hardwoods. Two great absorption decreases are visible at 1174 and 1139 cm^{-1}. The first decrease belongs to the asymmetric stretching of ether bond, while the second belongs to the symmetric stretching of ether bond, the aromatic C-H deformation and to the glucose ring vibration. These absorption decreases

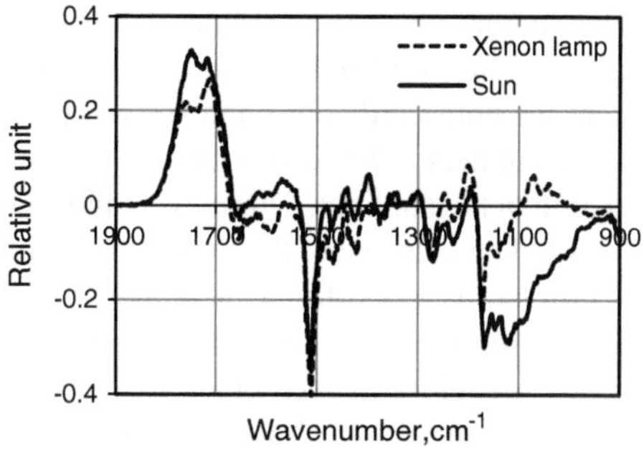

Fig. 2.53 Difference IR spectra of Spruce caused by 120-h light irradiation

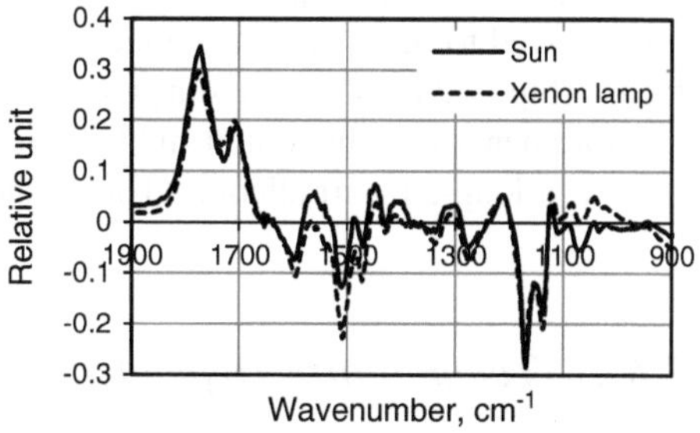

Fig. 2.54 Difference IR spectra of Black locust caused by 200-h light irradiation

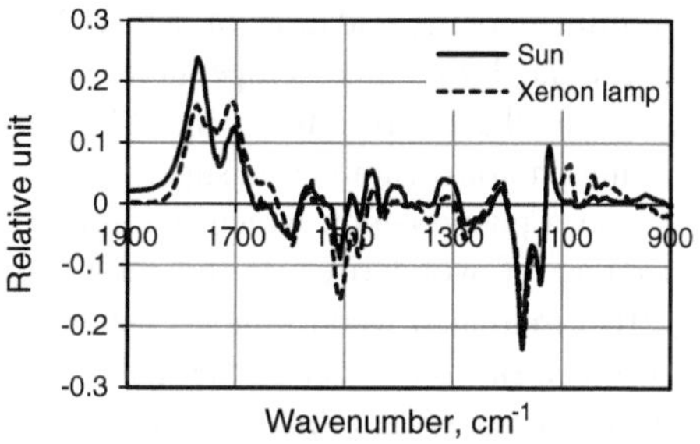

Fig. 2.55 Difference IR spectra of Black locust caused by 60-h light irradiation

indicate the ether splitting and the degradation of cellulose. (There is baseline difference between the difference spectra created by sunlight and xenon lamp in the 1400–1600 cm^{-1} wavenumber region.) The effect of the two irradiation types (sun and xenon lamp) differed in peaks of the 1680–1900 cm^{-1} wavenumber region. Usually, two peaks are developed in this region during the exposure of wood to UV radiation. In softwoods, these peaks are close to each other (Fig. 2.53); thus, the highly overlapped bands are sometimes visible as one broad band if the irradiation is strong enough or prolonged. For hardwoods, these bands are usually more separated (Figs. 2.54 and 2.55).

The absorption wavenumbers of the CO stretching in unconjugated ketones, carboxyl groups and lactones are between 1730 cm^{-1} and 1780 cm^{-1}, and the non-conjugated aliphatic carbonyls are absorbed between 1700 cm^{-1} and 1750 cm^{-1}.

The xenon light caused greater increase in the peak at 1716 cm^{-1} wavenumber than at 1748 cm^{-1} for Spruce. Sunlight produced similar change at 1716 cm^{-1} than xenon light for both hardwoods and softwoods, but some differences were found at higher wavenumbers. The xenon lamp irradiation produced smaller absorption increase here than the sunlight. These two bands are located at 1706 and 1770 cm^{-1} for Black locust. The increase of the two carbonyl bands had distinctive time dependence. The xenon lamp caused faster increase of the band around 1710 cm^{-1} than for the other band at higher wavenumber at the beginning of the treatment. The height of these two bands was close to each other after 200 h of irradiation for Spruce wood.

The change was similar for Back locust as well, but the height of the peaks occurred equal earlier, after 60 h of irradiation (Fig. 2.55). The difference between the effect of sunlight and xenon light decreased after this period. The effect of xenon light proved to be equal to the effect of sunlight after 200 h of irradiation on Black locust (Fig. 2.54). These findings strengthen our conclusion suggested by colour measurements that the xenon light irradiation can simulate the effects of sunlight only in a long-term irradiation (Tolvaj and Mitsui 2005).

2.6.2 *Temperature Dependence of Photodegradation*

The superposition effects of two or more influential parameters were hardly examined for degradation of wood. Matsuo et al. (2010), Matsuo et al. (2014) investigated the time–temperature superposition during thermal degradation of wood in the temperature range 90–180 °C. The effect of elevated temperature during photoirradiation (the superposition of light irradiation and thermal treatment) is also a phenomenon which is rarely investigated, even though the surface temperature of wood rises considerably during light irradiation. Persze and Tolvaj (2012) irradiated wood samples using mercury vapour lamp at 80 °C and at 30 °C to screen out the effect of thermal decomposition during photodegradation. Their results demonstrated that the same light irradiation resulted in considerably greater redness increase at 80 °C than at 30 °C. It was found that the extractive content has an important role in thermal decomposition during photodegradation.

The effects of consecutive photodegradation and thermal degradation (photoirradiation followed by thermal treatment or thermal treatment followed by photoirradiation) are widely investigated. Mitsui and his co-workers (Mitsui et al. 2001, 2004, 2005; Mitsui 2004) found that the combination of light and heat treatment (light irradiation was followed by heat treatment) creates even more discolourations than the separate light and heat exposures do. The colour change caused by heat treatment was much greater after light irradiation than without it. In these studies, the light irradiation was followed by dry and wet thermal treatment. The light irradiation produced small redness increase in dry condition, but the following thermal treatment created 8 times greater redness increase. The applied temperature was high (120–160 °C) in Mitsui's study. The change of IR spectrum showed different alterations. The intensity of carbonyl groups (that increased with light irradiation) decreased with wet heat treatment. The increment of two carbonyl sub-bands by light irradiation was similar, but the decrement of the carbonyl band at 1756 cm^{-1} was greater than at 1716 cm^{-1}, with heat treatment after light irradiation.

The other possibility, the photodegradation of previously thermally treated wood, was also investigated by several authors. Some papers explained that thermally treated wood had greater resistivity to photodegradation than non-treated ones. But others found that thermally treated wood undergoes the same photodegradation as untreated wood. More details can be found in Sect. 2.5.3.

The effect of low temperature was also investigated. Mitsui and Tsuchikawa (Mitsui and Tsuchikawa 2005) irradiated wood samples that were kept in a conditioning chamber set for −40 °C. They concluded that the measured degradation was much smaller at −40 °C than under usual laboratory condition.

In order to clarify the superposition effect of elevated temperatures during photodegradation of hardwood and softwood, samples were irradiated by mercury lamp at elevated temperatures (Tolvaj et al. in print). The investigated hardwood samples were as follows: Ash (*Fraxinus excelsior* L.) and Poplar (*Populus x euramericana Pannonia*), and the softwood samples were as follows: Scots pine (*P. sylvestris* L.) and Spruce (*P. abies* Karst.). The UV radiation of mercury vapour lamp was 80 % of the total (UV and visible) light emission (31 % UV-A, 24 % UV-B and 25 % UV-C). The total electric power of the applied double mercury lamps was 800 W, and the samples were located 64 cm from the lamps. Four different air temperatures 30, 80, 120 and 160 °C were used in the irradiation chamber. A series of samples was treated in the same chamber set for 30, 80, 120 and 160 °C but without light irradiation. The effect of pure thermal degradation was determined by this experiment. Total treatment time was 36 h in all cases. The colour of the wood specimens was measured before and after treatments. The exposures were interrupted after 6, 11, 16 and 36 h to measure the colour data. Measurements were carried out with a colorimeter (Konica-Minolta 2600d). The CIE L^* (lightness), a^* (redness) and b^* (yellowness) colour space data were calculated based on the D_{65} illuminant and 10° standard observer with a test-window diameter of 8 mm.

Figure 2.56 represents the superposition effects of photodegradation and thermal treatment, using simultaneous light irradiation and heat treatment. The temperature within the irradiation chamber including the samples was altered (30, 80, 120 and 160 °C), but the light irradiation always remained the same. The lightness value was increased at higher temperatures, but its effect between 30 and 120 °C was moderate. Moreover, the thermal effect of 30 °C is so small that it can be neglected and this circumstance allows the determination of the colour change caused by pure light irradiation (which is UV 30 °C in Fig. 2.56).

The only exception was the Scots pine having great lightness change (more than doubled) between 80 and 120 °C. The other 3 species also doubled the lightness change between 120 and 160 °C. There were no consequent differences in the lightness change among the species. Only Ash was a little more sensitive to the treatment than the others. It seems that the darkening increased exponentially by

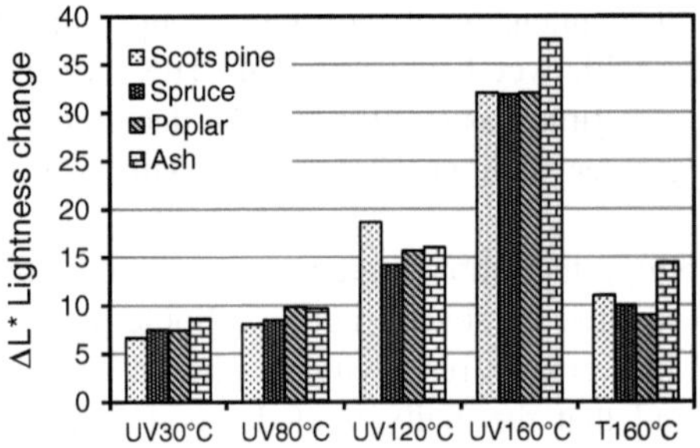

Fig. 2.56 Decrease of lightness (L^*) for the investigated wood species due to UV irradiation at different temperatures (*UV*) and only thermal treatment (*T*) at 160 °C. The duration of all treatments was 16 h

increasing temperature according to the Arrhenius law. (This statement will be clarified later.) The samples in the totally dark chamber showed considerably less darkening at 160 °C (last columns in Fig. 2.56) than the light irradiated samples at 160 °C.

The representation in Fig. 2.56 does not allow deeper insight into the lightness and colour hue variations as a function of influencing factors. Due to the simultaneous action of light irradiation and heat, it is interesting to examine first their separate effects.

The lightness and colour hue (b^*/a^*) variation of the investigated wood species due to pure light irradiation at ambient (30 °C) temperature are seen in Fig. 2.57. (Measurements showed that this low temperature did not cause any colour change during the 16 h in total darkness.) It is striking that in the first part of the treatment, all species became greyer and later, after 6 h treatment, the direction of change turned and the rate of change slowed. In 36 h of irradiation, the samples did not get back their initial colour hue.

The pure heat treatment has shown also remarkable anomalies at 120 and 160 °C temperatures (Fig. 2.58). At air temperatures of 120 °C, all samples moved towards

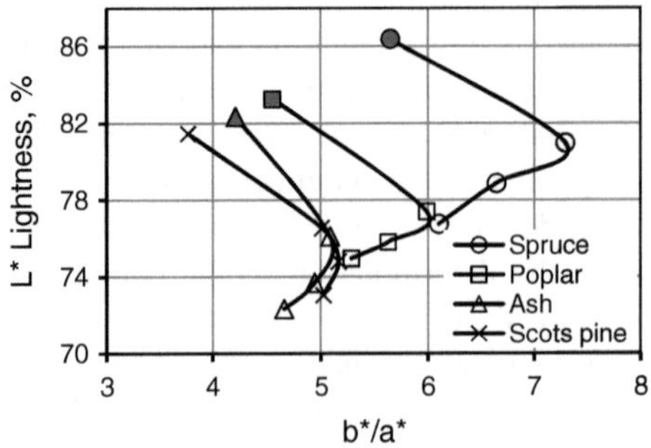

Fig. 2.57 Change of lightness and colour hue due to UV irradiation at 30 °C (*filled marks* means nonirradiated, followed by 6-, 16- and 36-h irradiations)

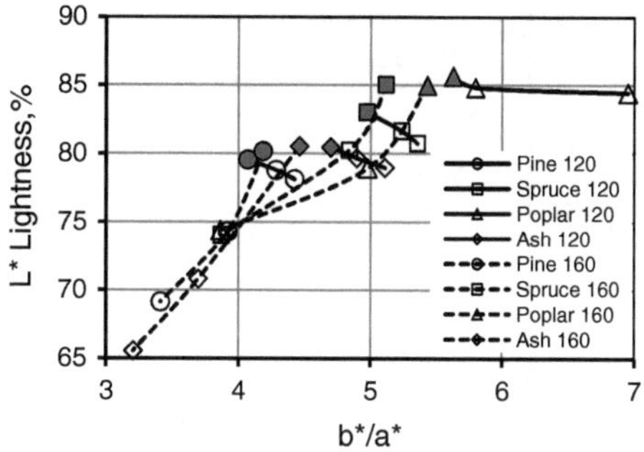

Fig. 2.58 Interaction between lightness and colour hue at heat treatment with different temperatures (*filled marks* means non-treated, followed by 6- and 16-h thermal treatments)

the grey colour hue, although the range of movement was different. Poplar produced the greatest hue change. The change in lightness was moderate in all cases. At 160 °C temperature, however, all samples became more brownish accompanied by high rate of lightness decrease. From these results, it follows that the temperature level, if it is acting alone, plays a definite role in the chemical reactions taking place during heat treatment. The IR measurement demonstrated that lignin did not suffer degradation during pure heat treatment. The thermal degradation of extractives has happened, and it was followed by the creation of quinones generating brown colour. The created brownish colour may partly be originated by the "caramellisation" of the sugar units produced by the degradation of the hemicelluloses as well.

The combined and simultaneous effects of light irradiation and heat treatment are demonstrated in Fig. 2.59. The measurement results show here, more or less, the usual picture; the samples became always more brownish and their lightness reduction is remarkable. An interesting feature is that the course of changes is always the same but naturally temperature dependent. The 160 °C generated much greater lightness and hue change than the 120 °C temperature.

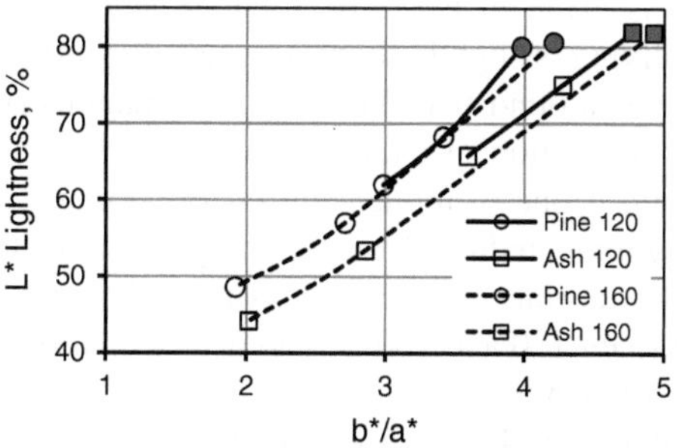

Fig. 2.59 Lightness and colour hue changes at simultaneous UV irradiation and heat treatment (*filled marks* means nonirradiated, followed by 6- and 16-h irradiations)

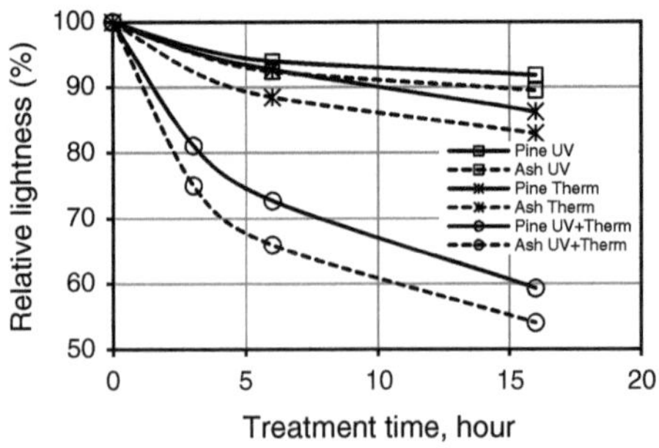

Fig. 2.60 Relative lightness and its components as a function of treatment duration for Pine and Ash (UV at 30 °C, Therm at 160 °C, UV+Therm at 160 °C)

In the following, it is interesting to examine the difference between the separate and simultaneous influences of light irradiation and heat treatment on lightness and colour hue. Knowing the separate effects from Figs. 2.57 and 2.58, they can be compared to those given in Fig. 2.59.

A more illustrative representation of the components and resultant of lightness variations is possible as a function of treatment duration (Fig. 2.60). It can clearly be seen that the sum of components is much less than the resultant due to the simultaneous effects. The simultaneous action of light irradiation and heat strengthen each other and a magnifying factor can be defined as follows:

$$M = \frac{\Delta L_{1+2}}{\Delta L_1 + \Delta L_2}$$

where .

ΔL_{1+2} is the lightness reduction due to combined actions of light irradiation and heat,

ΔL_1 is the lightness reduction due to light irradiation and

ΔL_2 is the lightness reduction due to heat treatment

This relationship is given in Fig. 2.61 for Pine and Ash at two different temperatures. The magnifying is slightly decreasing as a function of treatment duration, and its value fluctuates around two. Probably, the magnifying factor has a maximum somewhere as a function of temperature, for a given irradiation wavelength and intensity. Namely, if one of influencing factors dominates, then its prevailing action governs the lightness reduction and the magnifying factor tends to one. From these experiments, the accurate position of maximum cannot be given, but it is, with high probability, between 120 and 160 °C and nearer to 120 °C than to 160 °C (see the relative position of curves).

A comparison of colour hue is more complicated. The colour hue and its coordinates have a very complicated superposition law which is not known for the mixed colour of wood species. For example, the pure light irradiation and heat

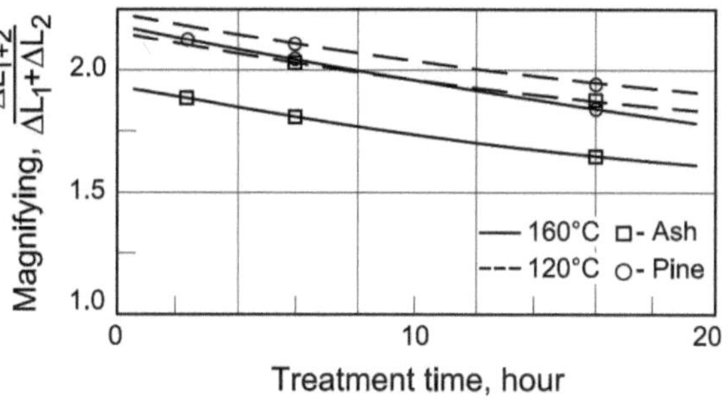

Fig. 2.61 Factor of magnifying as a function of treatment duration for two different temperatures

treatment at 120 °C both produce greyer colour hues, but their combined effect results in browner colour hues (see Figs. 2.57, 2.58 and 2.59).

A rough comparison for heat treatment and combined treatment at 160 °C is possible which is represented in Fig. 2.62. Keeping in mind the initial values and possible final values, the relationship is parabolic. Depending on the colouring materials in wood species, the correlation curves are slightly shifted to each other.

The redness change showed even greater differences between the sum of pure light and heat treatments and the simultaneous light and heat treatment than the lightness change (Fig. 2.63). Publications reported the softening and the condensation of lignin during thermal treatment above 100 °C (Popescu et al. 2013; Hill et al. 2006). It seems that the UV light irradiation is able to create much more quinones during the redistribution and condensation of lignin at 160 °C than at ambient temperatures. These quinone structures are partly responsible for the redness increase.

The yellow colour shift proved to be different compared to the change of lightness and redness. Figure 2.64 shows that there was a significant yellow coordinate increase even at 30 °C. This increase was 91 % for Poplar. (For comparison, the lightness decrease was 9 % and the redness increase was 55 % for Poplar at 30 °C.) The light irradiation at 80 °C produced similar yellowing as at 30 °C. The increase in yellowness is created mostly by the light degradation products of lignin. It means

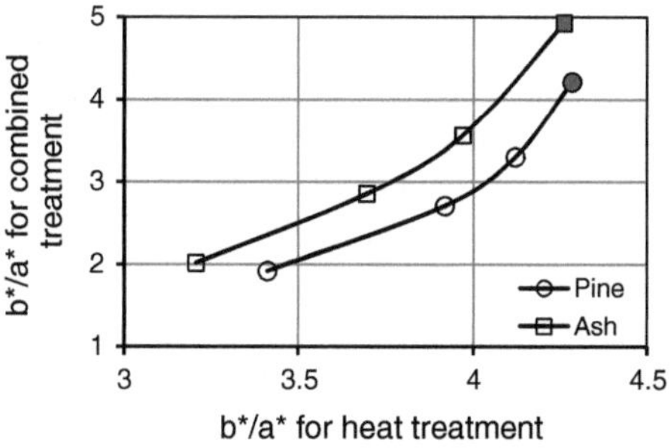

Fig. 2.62 A possible correlation between heat and combined treatments at 160 °C temperature (*filled marks* means non-treated, followed by 2-, 6- and 16-h treatments)

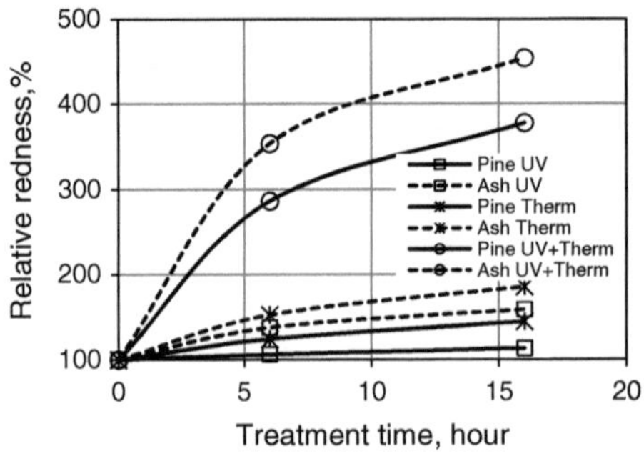

Fig. 2.63 Relative redness and its components as a function of treatment duration for Pine and Ash (UV at 30 °C, Therm at 160 °C, UV+Therm at 160 °C)

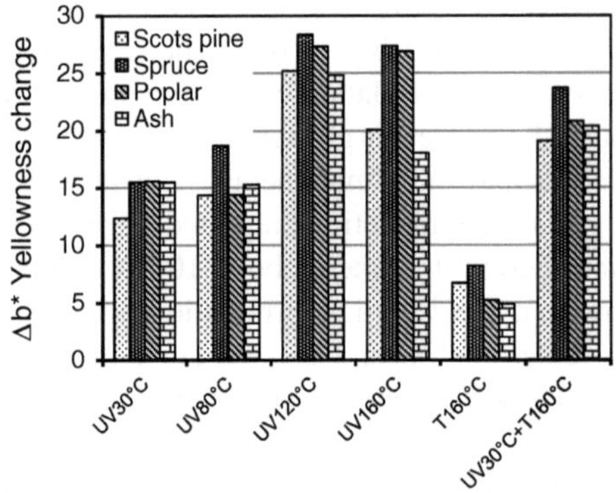

Fig. 2.64 Yellowness (*b**) change of the investigated wood species due to UV irradiation at different temperatures (*UV*) and only thermal treatment (*T*) at 160 °C. The duration of all treatments was 16 h

that the lignin degradation is not temperature dependent within the 30–80 °C temperature range. The only exception was the Spruce, producing considerably greater yellowing at 80 °C than at 30 °C. Spruce was more sensitive to yellowing at all applied treatments than the other species. It seems that the light degradation products of lignin are not enough stable at 160 °C. This high temperature can degrade the newly created chromophore compounds resulting in yellowness decrease. The pure thermal treatment at 160 °C produced much smaller yellowness change than the UV irradiation at the same temperature. Furthermore, this yellowness increase caused by the pure thermal treatment at 160 °C was much smaller than the yellowness increase created by the photodegradation at 30 and 80 °C. In contrast, the redness change was similar caused by both the photodegradation at 80 °C and the pure thermal treatment at 160 °C. These findings strengthen the fact that the red coordinate shift is created mostly by the degradation products of the extractives, while the yellowing is mostly due to the degradation products of lignin.

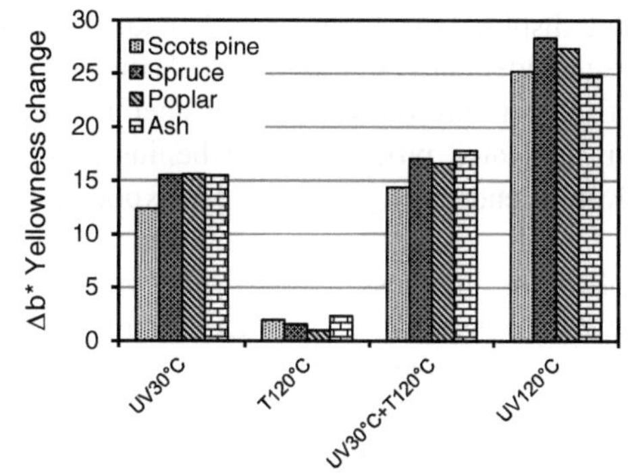

Fig. 2.65 Yellowness change of the investigated wood species due to UV irradiation at different temperatures (*UV*) and only thermal treatment (*T*) at 120 °C. The duration of all treatments was 16 h

The sum of the yellowness changes caused by two independent treatments (UV30 °C + T160 °C) was similar to the yellowness change caused by the simultaneous light irradiation and thermal treatment (UV160 °C). The small difference was positive for Spruce and Poplar and negative for Ash. This similarity does not mean that there is no multiplication effect from simultaneous treatments. Figure 2.64 clearly shows that there is a third effect at 160 °C as well. (The effect of 120 °C is greater than the effect of 160 °C.) Namely, this temperature is high enough to degrade new chemical products of photodegradation. This secondary degradation causes the decrease of yellowness at 160 °C.

At 120 °C, this secondary degradation probably does not occur, or its effect is small. The effects of 120 °C are visible in Fig. 2.65 showing the yellowness change of each wood species. The pure thermal treatment at 120 °C caused only a minor yellowness increase. It is much less than at 160 °C (see Fig. 2.64). The sum of the yellowness changes caused by the two separate treatments (UV30 °C + T120 °C) was considerably smaller than the yellowness change caused by the simultaneous light irradiation and heat treatment (UV120 °C).

The Arrhenius equation (2.33) relates the rate of a chemical reaction k to temperature T, and it includes the activation energy. The equation is simple if the activation energy (E_a) is constant. The Arrhenius equation can be given in the form:

$$k = A\mathrm{e}^{-E_a/(RT)} \tag{2.33}$$

where R is the universal gas constant.

The former form can be written equivalently taking logarithm natural as follows:

$$\ln k = \frac{-E_a}{R}\frac{1}{T} + \ln A$$

An Arrhenius plot displays the logarithm of a kinetic constant (*k*, ordinate) plotted against inverse temperature (1/*T*, abscissa). Arrhenius plots are often used to analyse the effect of temperature on the rates of chemical reactions. For a single rate-limited thermally activated process, an Arrhenius plot gives a straight line, from which the activation energy (E_a) and the pre-exponential factor (*A*) can both be determined.

The Arrhenius plots for lightness, redness and yellowness are presented in Figs. 2.66, 2.67 and 2.68, respectively. The data of one selected species are presented in all cases, because the other examined species showed similar behaviours. All Arrhenius plots have a breaking point close to 100 °C. The logarithm of lightness decreased with increasing temperature (Fig. 2.66) shown here by the data of Ash. The tendency of decrease was similar for both dry thermal treatment and combined UV and thermal treatment as well. The intensity of decrease was much greater for combined UV and thermal treatment than for dry thermal treatment. The trend lines show a breaking point. It means that the lightness change is determined by more than one temperature-dependent chemical changes above 100 °C. The hemicelluloses are the most temperature sensitive molecules among the main substance molecules of wood. The thermal degradation of hemicelluloses can be one of the additional chemical changes above 100 °C. The other effect can be the creation of quinones during photodegradation at elevated temperatures. Popescu et al. (2013) reported the softening and the condensation of lignin during thermal treatment at 140 °C. Probably, the UV light irradiation is able to create many more quinones during the redistribution and condensation of lignin at 160 °C than at ambient temperatures. Kubovsky and Kacík (2014) irradiated wood samples by infrared light which was produced by CO_2 laser. They found that the high input of energy accelerates the interaction of the functional groups resulting in the subsequent condensation of lignin. Condensation occurs at the α-carbon to form condensed structures, mainly of the diphenylmethane type. This type of structures oxidises easily and forms quinones which form strong hydrogen bonds with hydroquinone and phenol. Similar quinone structures can be created by UV light at elevated temperature above 100 °C. These quinone structures can be partly responsible for the increase in darkening and for the increase in redness.

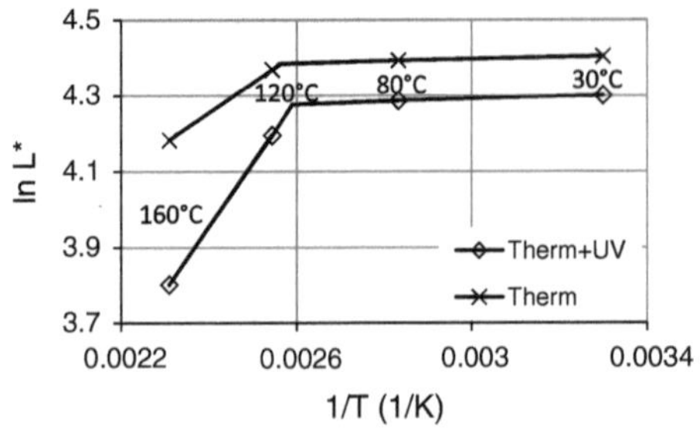

Fig. 2.66 The Arrhenius plots of lightness for Ash samples obtained by UV irradiation (*Therm +UV*) and dry thermal treatment (*Therm*) at the indicated temperatures

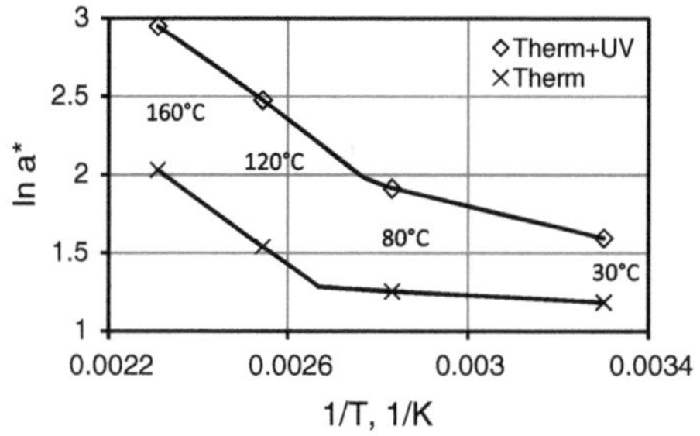

Fig. 2.67 The Arrhenius plots of redness for Spruce samples obtained by UV irradiation (*Therm +UV*) and dry thermal treatment (*Therm*) at the indicated temperatures

The Arrhenius plots of redness are shown in Fig. 2.67 presented here by the data of Spruce. The logarithm of redness increased with rising temperature. The increase was more rapid for combined UV and thermal treatment than for pure dry thermal treatment. The Arrhenius plots are close to linear in both cases. This is because the redness change is determined mostly by the degradation products of the extractives. Having a closer look at the data of Fig. 2.67, a breaking point can be observed between 90 and 100 °C. Below this point, the redness increase was slightly slower for both type of treatment than above this point. This breaking point is located at 95 °C in case of Black locust for the colour modification caused by steaming. The breaking point represents the complexity of the process. The degradation of hemicelluloses becomes visible above this temperature beside the degradation of the extractives. The quinone structures can be partly responsible for the redness increase.

The Arrhenius equation indicates that at any temperature, there will always be some reaction, though it may be immeasurably slow if the temperature is low enough. Even this slow change becomes visible if the time period is long enough. This phenomenon is visible on indoor claddings made of Spruce. The originally yellowish surface becomes more and more reddish during the years even at 20–25 °C.

The Arrhenius plots of yellowness are shown in Fig. 2.68 presented by the data of Scots pine. The logarithm of yellowness increased with rising temperature. The tendency of change was completely different for combined UV and thermal

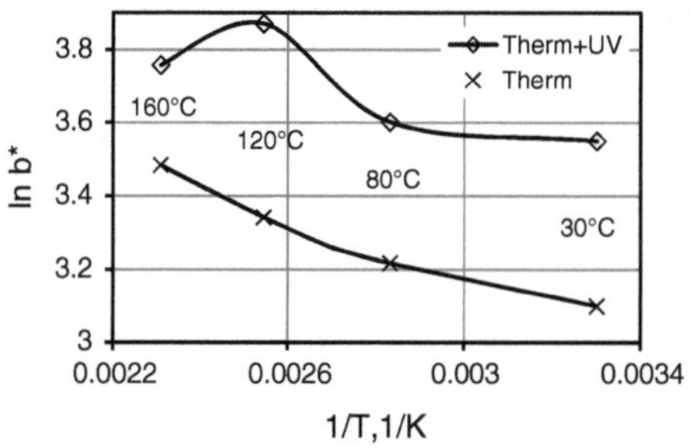

Fig. 2.68 The Arrhenius plots of yellowness for Scots pine samples obtained by UV irradiation (*Therm+UV*) and dry thermal treatment (*Therm*) at the indicated temperatures

treatment compared with pure dry thermal treatment. The pure thermal treatment created continuous, almost linear change with rising temperature. It represents that the thermal degradation of extractives follows the Arrhenius law. There is a hardly visible breaking point here as well, as in Fig. 2.67. The photodegradation of wood at elevated temperature is a highly complex process. That is why the Arrhenius plot is not a straight line. The yellowness change caused by the photodegradation is created mainly by the degradation products of lignin. The lignin degradation is not temperature dependent below 80 °C as clearly shown in Fig. 2.64. Only a small yellowness decrease was detected for Poplar and Ash and a little increase for the conifers between 30 and 80 °C. Above 80 °C, the yellowness increase was rapid, but it is visible only up to 120 °C. Above 120 °C, the degradation products of lignin undergo further thermal degradation reducing the yellowness.

It can be concluded that UV light irradiation produces much greater colour change at elevated temperatures above 80 °C than at ambient temperature. All 3 colour parameters (L^*, a^*, b^*) have shown this magnification effect for elevated temperatures. The colour change caused by UV light irradiation at 30 °C (effect of UV light) was added to the colour change caused by pure thermal treatment at 160 °C. The sum of the colour changes caused by these two separate treatments was considerably smaller than the colour change caused by the simultaneous light irradiation and thermal treatment at 160 °C. The thermal effect during photodegradation was not only the simple addition of two effects, but also the elevated temperature multiplied the effect of photodegradation. The greatest difference was revealed in the redness change.

The Arrhenius plots of all three colour coordinates had a breaking point close to 100 °C, showing that above this temperature the chemical changes are more complex than below this value. The redness changes could be described by the Arrhenius law, indicating that both light and thermal degradations of the extractives are exponential functions of the applied temperatures. The yellowness change caused by pure thermal treatment followed the Arrhenius law; however, the simultaneous photodegradation and thermal degradation did not follow the Arrhenius law.

2.6.3 Air Humidity Dependence of Photodegradation

To test the humidity dependence of photodegradation, Beech and Spruce samples were put into a closed quartz tube (Tolvaj et al. 2015). A water bath under the samples ensured 100 % relative air humidity within the tube (wet condition). The temperature inside the tube was measured by thermocouple. For comparison, samples were closed into a similar dry tube (without water bath). Mercury vapour lamp was used for UV irradiation. The total electric power of the applied double mercury lamps were 800 W, and the samples were located 64 cm from the lamps. The temperature was 32 °C for the first series and 53 °C for the second series within the wet tube during the irradiation. (The temperature within the irradiation chamber was

Fig. 2.69 Lightness change of Spruce (*S*) samples due to UV light irradiation under wet (*W*) and dry (*D*) conditions at the indicated temperatures

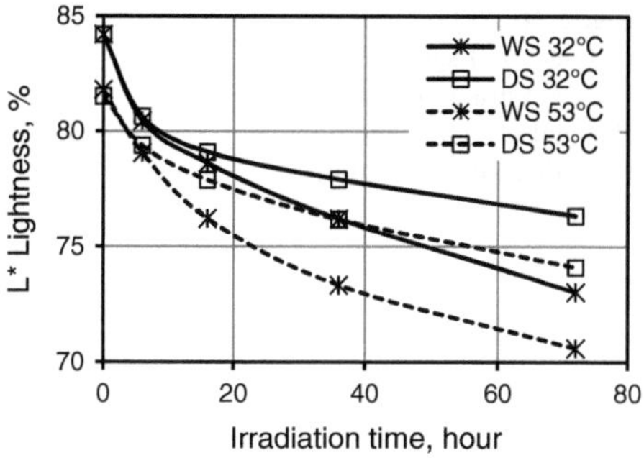

set to 30 and 50 °C.) The irradiation was interrupted after 6, 16, 36 and 72 h for colour and IR measurements. Five replicates were cut from the same board for both species with a dimension of 30 × 10 × 5 mm. The initial colour data of samples were slightly different used at 32 and 53 °C. The initial moisture content of the samples was 12 ± 1 %, and it increased up to 16 ± 2 % during the treatment at 50 °C.

The usual decrease in lightness was observed in all cases and for both examined species (Fig. 2.69). The rapid decrease in lightness at the beginning was followed by a moderate decrease. The wet condition caused considerably greater darkening than the dry condition at both temperatures. The treatment at higher temperature produced greater differences in lightness between wet and dry condition in the 10–60 h of irradiation.

The wet condition generated greater red colour change than the dry condition for all examined species at both temperatures (Fig. 2.70). The elevated temperature (53 °C) produced more rapid redness increase at the beginning of the treatment than the lower one (32 °C). After 36 h, the trend lines were close to parallel.

The yellow colour change of the samples was similar for both examined species at both temperatures (Fig. 2.71); nevertheless, the wet condition produced a slightly greater change than the dry condition. The treatments generated similar differences at both applied temperatures.

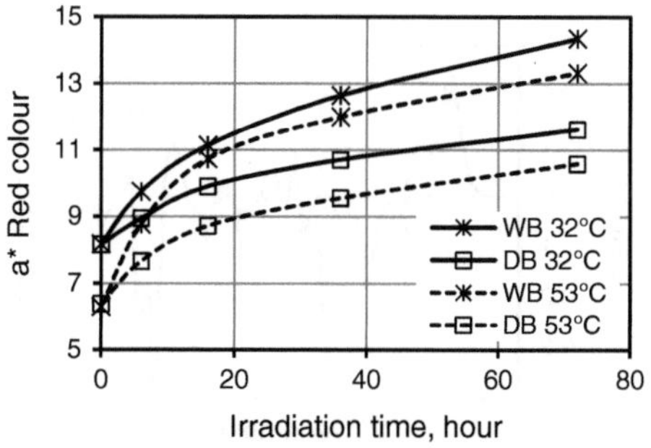

Fig. 2.70 Change in the redness of Beech (*B*) samples due to UV light irradiation under wet (*W*) and dry (*D*) conditions at the indicated temperatures

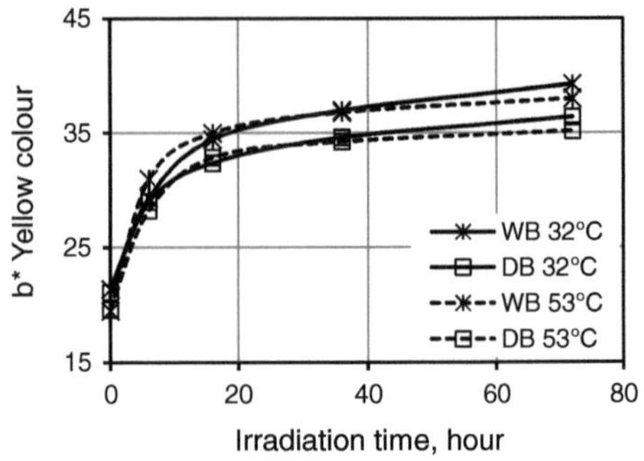

Fig. 2.71 Yellowness change of Beech (*B*) samples due to UV light irradiation under wet (*W*) and dry (*D*) conditions at the indicated temperatures

The wet condition produced considerably greater colour change than the dry condition. This finding is especially important as the air humidity in tropical countries is much higher than in continental climate. Consequently, results of different outdoor weathering tests are not comparable unless the air humidity data are carefully monitored.

Figures 2.72, 2.73 and 2.74 represent the difference IR absorption data of the samples produced by 36-h UV treatment. The UV light induced lignin degradation is clearly shown by the absorption decrease at 1507 cm^{-1} for both wood species.

The wet condition generated greater lignin degradation than the dry condition, i.e. the difference was greater at 53 °C than at 32 °C (Figs. 2.72 and 2.73). This finding is in good correspondence with the colour change data (Figs. 2.69, 2.70 and 2.71). The degradation products of lignin undergo rapid oxidation creating carbonyl groups. These carbonyl groups absorb the IR radiation between 1680 and 1820 cm^{-1}. According to the lignin degradation, the wet condition should generate greater absorption increase in the carbonyl region than the dry condition. It happened only to a small extent. The peak at 1706 cm^{-1} was a little higher in wet condition than in dry condition. The behaviour of the other peak at 1767 cm^{-1} was the opposite. The deviation was even greater at 53 °C. It seems that some parts of the carbonyl groups suffer further degradation.

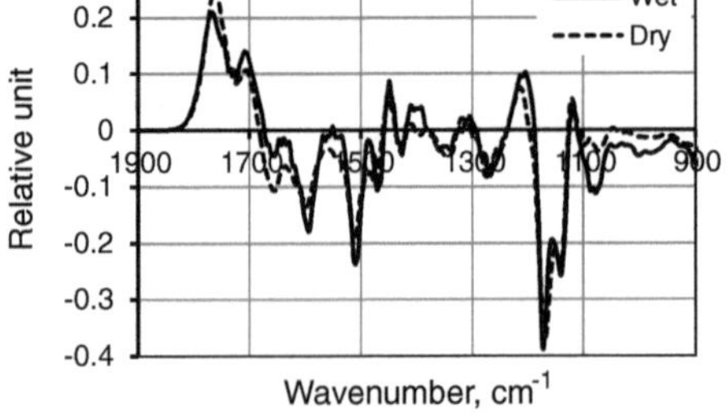

Fig. 2.72 Difference IR spectra of Beech samples caused by 36-h UV treatment at 32 °C under wet and dry conditions

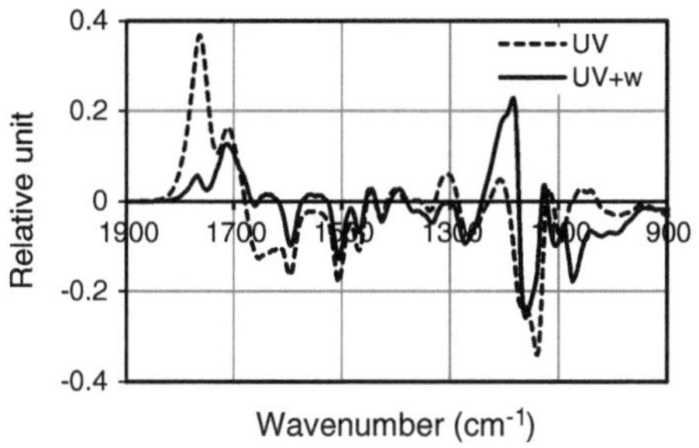

Fig. 2.73 The difference IR spectra of Beech samples caused by 36-h UV treatment at 53 °C under wet and dry conditions

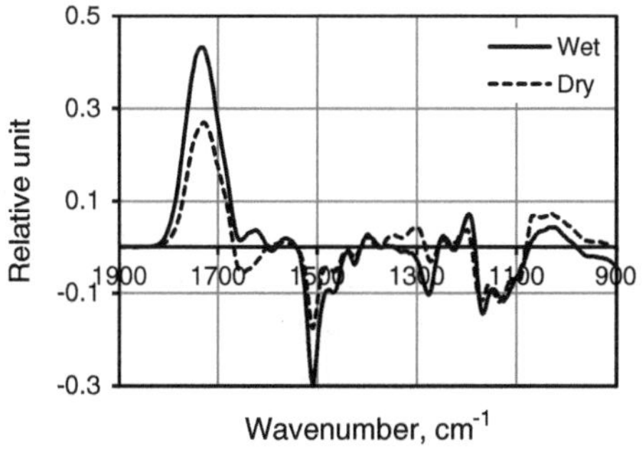

Fig. 2.74 Difference IR spectra of Spruce samples caused by 36-h UV treatment at 53 °C under wet and dry conditions

This is the "leaching" effect of the water molecules. This effect is discussed in details in the Sect. 2.6.4. Liquid water is able to leach out those chemical compounds containing carbonyl groups, which absorb around 1767 cm^{-1}. It seems that water molecules being in the air are able to remove carbonyl groups from the surface of the wood as well. The elevated temperature (53 °C) produced a more intensive removal than the lower (32 °C). The reason is that the vapour molecules move faster at higher temperature.

There are both absorption increases and decreases in the 1100–1300 cm^{-1} wavenumber region (Fig. 2.73). The absorption increases are mainly generated by the irradiation under wet condition. These are not real absorption increases. The increased roughness elevated the spectrum in this region. (Detailed explanation of the phenomenon can be found in the end part of Sect. 2.5.1.) This lifting effect reduced the intensity of the negative peak at 1157 cm^{-1}. The intensity of this negative peak should increase with elapsed irradiation time. But this intensity was −0.18, −0.21 and −0.22 after 16-, 36- and 72-h UV irradiation, respectively. The roughness increase was determined by optical method described in Sect. 2.5.1. The roughness values (Eq. 2.31) were 1.31, 1.34, 1.38 and 1.4 after 6-, 16-, 36- and 72-h irradiation, respectively.

Spruce samples (Fig. 2.74) have shown partly different behaviour than Beech. The lignin degradation at 1506 cm^{-1} was similar, but the absorption increase of the carbonyl groups differed considerably. The wet condition generated greater absorption increase of the carbonyl groups than the dry condition. This is opposite compared to what we have seen in the case of Beech. The 36-h UV treatment in wet condition produced 1.6 times greater absorption increase than in dry condition. This multiplication factor was 1.7 and 1.4 with 16 and 72 h treatment, respectively. The decrease of this multiplication factor indicates the existence of removal effect. The main difference between the IR spectra of Beech and Spruce is that there is no increasing absorption peak at 1760 cm^{-1} for Spruce (see Fig. 2.74). This is the band which was reduced by the leaching effect in the case of Beech.

Our results show that the presence of water vapour intensifies the degradation of lignin. Some parts of carbonyl groups (generated by the lignin degradation) are not stable, and the presence of vapour decreases the number of this chemical group. The increasing temperature amplifies this removing effect.

Nowadays, many papers represent the colour change only by the total colour change ΔE^*. However, the ΔE^* is not always reliable by itself. In some cases, ΔE^* values do not match to the visual observations. Human eyes are highly sensitive to the changes of hue determined by the alterations of a^* and b^*. The total colour change is mainly determined by the change of the dominant component. In the photodegradation of wood, the yellowing is the dominant change since the red colour shift is usually much smaller. These statements are well demonstrated in Fig. 2.75. This figure represents the colour changes of Beech caused by UV irradiation at 53 °C under wet condition (100 % air humidity).

It is clearly seen that the total colour change ΔE^* is mainly determined by the yellowness change. The total colour change was calculated without the redness change ($\Delta E'$) as well. The difference between these two curves (ΔE^* and $\Delta E'$) is negligible. It suggests that the effect of redness change is also negligible for the total colour change. This redness change is large enough to be visible to the naked eye. This example shows the disadvantage of using total colour change as one single indicator.

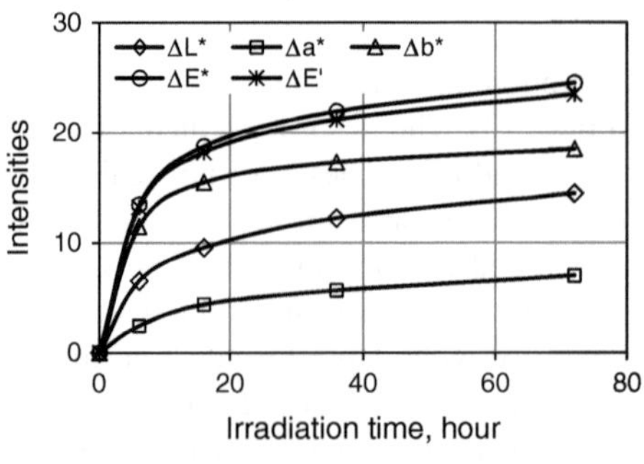

Fig. 2.75 The change of colour parameters for Beech caused by UV light irradiation under wet condition at 53 °C (ΔL is the absolute value of the lightness change and $\Delta E'$ is the total colour change calculated without the redness change)

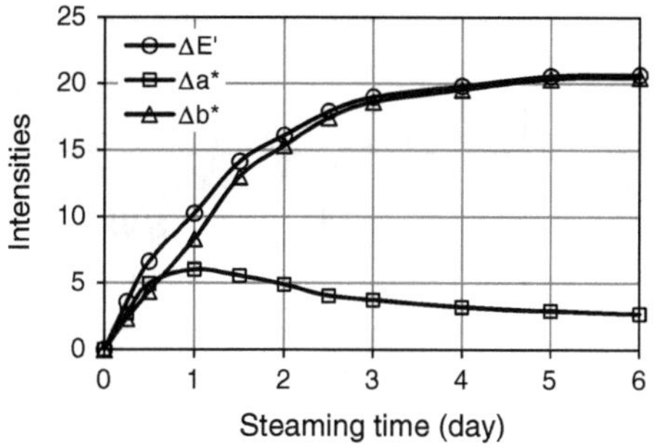

Fig. 2.76 The change of colour parameters for Black locust caused by steaming at 115 °C (Δb^* is the absolute value of the yellowness change and $\Delta E'$ is the total colour change calculated without the lightness change.)

As the changing tendencies of all three colour coordinates are similar, their ratio within the total colour change is not visible in Fig. 2.75. Another example is given in Fig. 2.76. The colour of Black locust can be considerably modified by steaming. During the process, its dominant unattractive yellow colour decreases, while the redness increases during the first day but decreases during the remaining period of the treatment. The total colour change ($\Delta E'$), calculated without the lightness change, does not follow the redness change. However, it is determined by the more intensive yellowness change.

These examples prove the weakness of the total colour change as one single parameter for monitoring colour alterations. The total colour change is determined mainly by the dominant changes and it disregards the small but visible alterations. Therefore, the three colour coordinates supply much more and detailed information than the total colour change.

2.6.4 Effect of Water Leaching

Beside sun radiation, the leaching effect of the rain is the second harmful factor effecting the degradation of outdoor wood. These effects were simulated by the following experiments. Samples were irradiated by mercury lamp and then plunged into distilled water for 1 day. Figure 2.77 represents the effect of the 1-day UV treatment of Spruce earlywood followed by 1-day of leaching in distilled water. The same initial spectrum was subtracted from the treated spectrum to calculate the difference spectra in both cases. The first difference spectrum presents the effect of UV treatment alone, while the second one shows the effect of UV treatment and the subsequent water leaching together. The UV light induced lignin degradation represented by the absorption decrease at 1507 cm^{-1} is clearly visible (Fig. 2.77). The water leaching did not affect this peak. The UV treatment produced great absorption increase in the unconjugated carbonyl region (1670–1820 cm^{-1}).

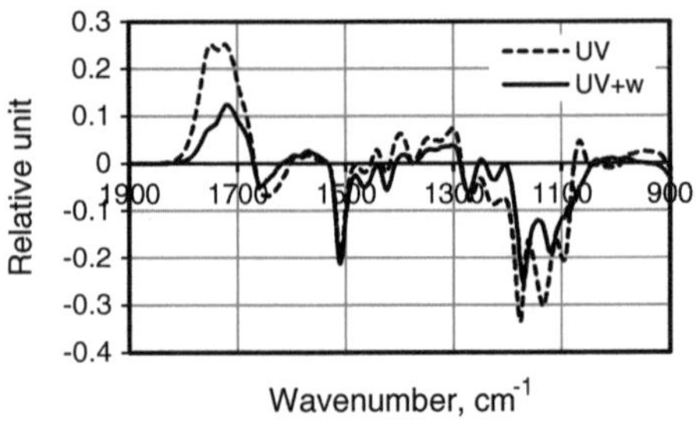

Fig. 2.77 Absorption change of Spruce earlywood after 1-day UV treatment (*UV*) and 1-day UV treatment followed by 1-day water leaching (*UV+w*) (the initial spectrum was the reference in both cases)

Two peaks arose close to each other during the one-day treatment. (The longer treatment unites these bands into one band with a maximum at 1720 cm^{-1}.) The intensity of this band was reduced considerably by leaching. The intensity of both peaks (1713, 1736 cm^{-1}) decreased but to a different extent. The peak at 1713 cm^{-1} lost half of its intensity, while the other peak at 1736 cm^{-1} almost disappeared and remained only as a shoulder. The little shoulder at 1680 cm^{-1} became clearly visible after leaching, strengthening the claim that three peaks arose in the unconjugated region as the result of photodegradation. This third peak was hardly affected by water leaching. The UV treatment generated absorption decrease in the ether bond region (1070–1200 cm^{-1}). Three negative bands occurred representing the splitting of the ether bond in the cellulose chain, in the pyranose ring and in the aromatic system.

Examining the photodegradation and leaching properties of Beech (Fig. 2.78), similar changes were found compared to Spruce earlywood. The lignin content was not affected by the water leaching. It is exemplified by the negative peak at 1596 and 1508 cm^{-1}. The water leaching does not change the intensity of these lignin peaks. Here, the apparent difference is visible because of baseline shift. The UV radiation degraded the conjugated carbonyl groups in the extractives presented here

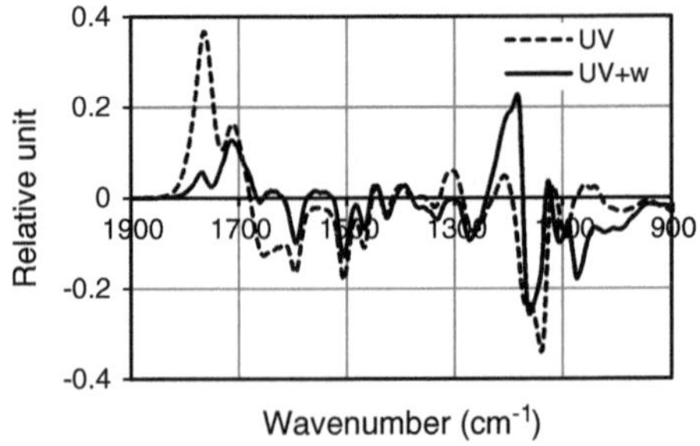

Fig. 2.78 Absorption change of Beech after 1-day UV treatment (*UV*) and 1-day UV treatment followed by 1-day water leaching (*UV+w*) (the initial spectrum was the reference in both cases)

by the absorption decrease around 1650 cm^{-1}. The subsequent water leaching eliminated this negative peak. An absorption band of water is located here as well. Probably, the moisture content of the samples after leaching and drying was not exactly the same as the initial moisture content of the untreated samples. The absorption increase of the water overlapped the absorption decrease of the conjugated carbonyls creating zero difference. This effect is visible in Figs. 2.72 and 2.74 but in different quantity.

Two peaks were grown up in the non-conjugated carbonyl region (1670–1820 cm^{-1}) as a result of the UV light induced lignin degradation. The stability of these two peaks is not equal. The peak at 1706 cm^{-1} was hardly affected by the water leaching, while the other peak at 1767 cm^{-1} was mostly removed by the water leaching. This erosion of the surface allows the photodegradation of the next layer. As the earlywood is more degradable than the latewood, the outdoor surface of wood will become rougher during the years.

The interpretation of the changes in the ether band region (1100–1200 cm^{-1}) is difficult. Although the absorption intensities were calculated from the diffuse reflectance spectra using the K-M theory, this theory does not work well in this region where the absorption of wood is strong (see Sect. 2.5.1). The increasing roughness lifts up the difference spectrum confusing the real changes. The water leaching (and drying) produces roughness increase too (see later in Sect. 4.10). The swelling lifts up the cut end of the fibres, and the drying does not bring them back creating a microlevel increase in roughness. The lifting effect is more visible in Figs. 2.79 and 2.80.

To monitor the effect of repeated rainfall, the samples were irradiated by mercury lamp (first time 1 day) followed by water leaching (always 1 day). The repeated light irradiations were for 2 days followed by 1 day water leaching. For comparison, continuous light irradiation (without leaching) was also carried out. Results of light irradiation are presented here until the 7th day (4 cycles). Small samples (30 × 10 × 5 mm) were dried at room temperature for a week after water leaching for IR measurement. The changes were always compared to the untreated

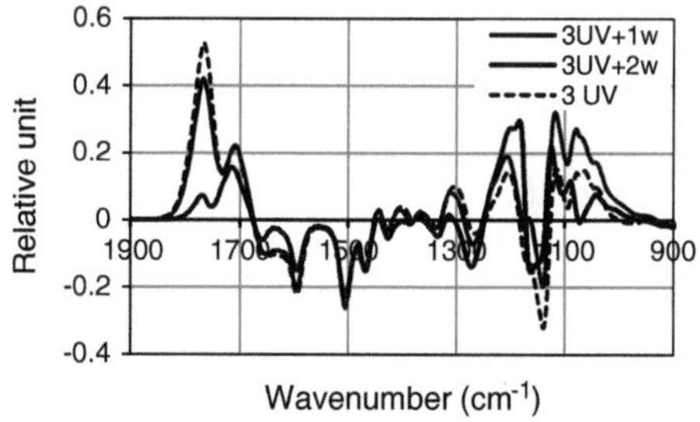

Fig. 2.79 Absorption change of Beech after different treatments (the reference is the initial spectrum in all cases.). *Legend* 3 UV = 3-day UV treatment; 3UV+1w = 1-day UV irradiation, 1-day leaching followed by 2-day UV irradiation; 3UV+2w = 1-day UV irradiation, 1-day leaching, 2-day UV irradiation followed by 1-day leaching

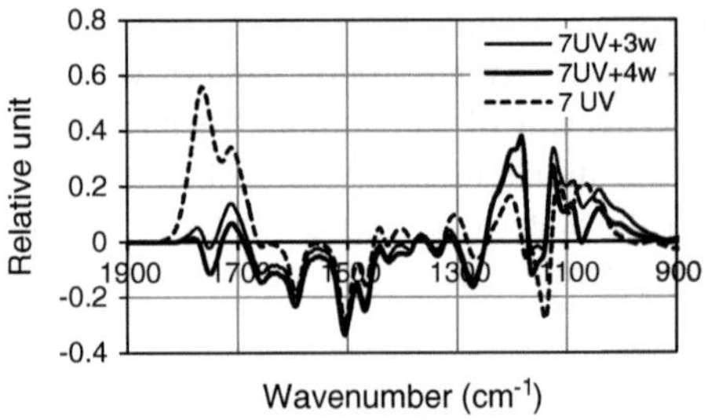

Fig. 2.80 Absorption change of Beech after different treatments (the reference is the initial spectrum in all cases). *Legend* 7 UV = 7-day UV treatment; 7UV+3w = 1-day UV irradiation and then 1-day leaching followed two times by the combination of 2-day UV irradiation and 1-day leaching and then finished with 2-day UV irradiation; 7UV+4w = 1-day UV irradiation and then 1-day leaching followed three times by the combination of 2-day UV irradiation and 1-day leaching

samples independently of the numbers and type of treatment. Always the initial spectrum was subtracted from the treated one to create the difference spectrum.

The lignin degradation (presented by the absorption decrease at 1505 cm^{-1}) was similar for both continuous UV treatment and UV treatment interrupted by water leaching in the case of all cycles. This experience confirms that the lignin degradation was not affected by water leaching. The results of the second series of treatment are presented in Fig. 2.79. The 1 + 2-day light irradiation produced similar but only a little greater changes than the 1-day light irradiation. The 3-day continuous UV treatment created a little greater absorption increase at 1767 cm^{-1} than the treatment interrupted by leaching. The only important difference between Fig. 2.78 and Fig. 2.79 is that the K-M transformation lifts up the difference spectrum between 1070 and 1200 cm^{-1} because of the increasing surface roughness of the samples. The lignin degradation was continuous with increasing light irradiation for both UV treatment types (with and without leaching). The absorption of carbonyl groups at 1767 and 1706 cm^{-1} increased continuously only for continuous UV treatment.

The UV treatment interrupted by leaching created a declining rate of production of carbonyl groups. The intensity of the peak at 1706 cm^{-1} decreased slowly after the second cycles, but the intensity of peak at 1767 cm^{-1} decreased rapidly. This phenomenon can be interpreted by the lack of the lignin molecules on the surface of the sample. The reduction of the absorption increase at 1767 cm^{-1} is clearly visible in Fig. 2.80. This absorption increase was negligible even before the water leaching during the fourth cycles (7UV+3w). The next leaching reduced this small positive peak as well. The fourth day of leaching created a negative peak at 1750 cm^{-1}.

This absorption decrease did not originate from the carbonyls created by photodegradation, but displays the degradation of carbonyl groups being originally in the untreated wood. This negative peak was never found before. Our suggestion was that the water leaching generated this negative peak independently from photodegradation. To clarify the reason for the absorption decrease at 1750 cm^{-1}, wood samples were put into distilled water to find out the effects of pure water leaching. After a 3-day leaching, a small absorption decrease at 1750 cm^{-1} was detected, and it became clearly visible after the fourth day of leaching. This finding proves that the rain is able to leach out not only the carbonyl groups created by photodegradation but also some type of carbonyls originally present in the wood.

2.7 Hardness of Wood Surfaces

The durability and mechanical resistance of wood surfaces against different kind of loadings (static, dynamic, concentrated, frictional, etc.) considerably depend on their hardness. Hardness measurements are widely used in the metal industry, and several standardised methods are known. Due to the structural properties of wood, hemispherical indenters are more suitable for hardness determination in the wood industry. The diameter of indenters varies generally between 25 and 35 mm, and the earlier users have often used inch dimensions (e.g. 5/4 inch).

Hardness measurement of wood materials have been begun 100 years ago thanks to the pioneering work of Janka (Janka and Wien 1912; Janka et al. 1915) who proposed and introduced the application of hemispherical indenter in 1912. The hemispherical shape also allows a more accurate mathematical treatment although the combined elastic–plastic deformations taking place in the wood material gives rise to some difficulties in the evaluation of the measurement results.

The mode of loading on the surface by spherical indenters may be static or dynamic (Fig. 2.81). With static loading, the indenter is pushed into the surface with low velocity, and the force is recorded as the function of displacement. If the sinkage of indenter is prescribed, then the measurement of the maximum force is also sufficient.

Performing dynamical measurements, a metal sphere with a mass m is dropped onto the surface from given height h having the kinetic energy of:

$$E = mgh = \frac{m}{2} v_0^2$$

The hardness is a secondary mechanical property due to its complicated multiaxial stress state, the anisotropy of wood and permanent deformations. Its numerical measure considerably depends on the measurement technique used.

The calculations are based on a simplified engineering mechanics (Hertz and Boussinesq problems) supposing small deformations. In spite of these simplifications, however, the hardness of wood is a useful mechanical property for comparing different wood species concerning their surface stability and durability.

Using the static measurement method, a spherical indenter with a radius R is pressed into the wood surface and, according to the Hertz theory, the force F is calculated with the following relationship:

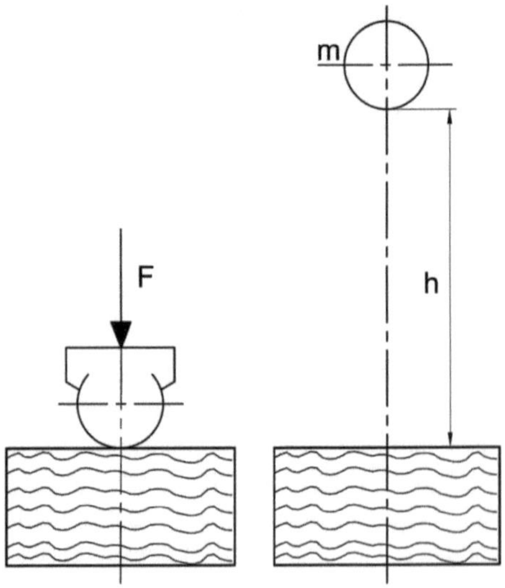

Fig. 2.81 Static and dynamic measurement method for hardness

$$F = \frac{4}{3}\frac{E}{1-\nu^2}\sqrt{R}z^{3/2} = Bz^{3/2} \quad (2.34)$$

where E and ν are the modulus of elasticity and Poisson's ratio, respectively, and z is the deformation under the indenter.

In Hungary, a spherical ball of 5/4 inch dia. has long been used with a fixed 2-mm deformation. In this case, the hardness H_w is characterised by the material property $E/(1-\nu^2)$, and it is given by:

$$H_w = \frac{E}{1-\nu^2} = \frac{F}{\frac{4}{3}\sqrt{R}z^{3/2}} = \frac{F}{15} \quad \mathrm{N/mm^2} \quad (2.35)$$

where F is the force for the given deformation of $z = 2$ mm.

From Eq. (2.35), the reference surface area under the indenter is given as follows:

$$A = \frac{4}{3}\sqrt{R}z^{3/2}$$

Sometimes, the contact surface of the indenter as a reference area was proposed. The surface of the spherical calotte (A_c) is given by the equation:

$$A_c = 2\pi Rz$$

which fundamentally differs from the above reference area equation and indicates that the use of this latter surface area is theoretically not correct.

The dynamic measurement method generally uses a metal sphere with the mass *m,* and it is dropped onto the surface from given height *h*. Using the energy equation, we may write as follows:

$$mgh = \int_0^{z_0} F(z)\mathrm{d}z$$

where the force function is given by Eq. (2.34). Integrating the above equation yields as follows:

$$H_\mathrm{d} = \frac{E}{1-\nu^2} = \frac{mgh}{\frac{8}{15}\sqrt{R}z_0^{5/2}} \quad \mathrm{J/cm^3 or\ N/cm^2} \tag{2.36}$$

Now, the denominator has the dimension of cm^3, but this volume again is not identical with the volume of the spherical calotte. The maximum deformation z_0 can be obtained from the measured radius of the imprint (Fig. 2.82) by the simple relation:

$$z_0 = R - \sqrt{R^2 - r^2}$$

The maximum deformation can also be expressed from Eq. (2.36) as follows:

$$z_0 = \left(\frac{2.5\,mgh}{B}\right)^{0.4} \tag{2.37}$$

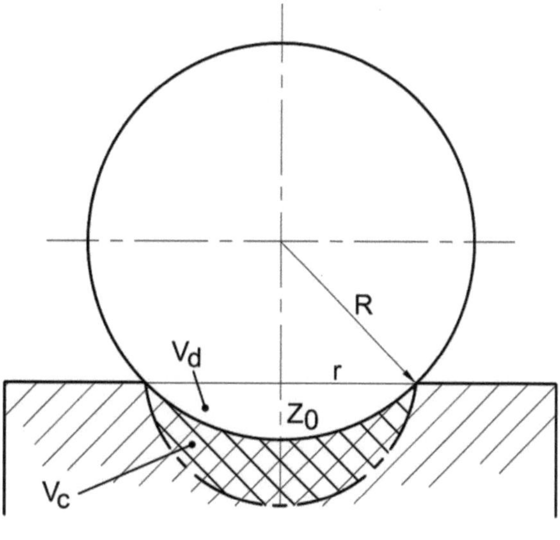

Fig. 2.82 Deformations under spherical indenter

Table 2.3 Calculated hardness values

	r mm	z_0 (mm)	H_d (N/mm^2)	V_d (mm^3)	F_{max} (N)	T (s)	v (m/s)
Ash	5	0.57	2835	22.5	7674	12.76×10^{-4}	2.1
Spruce	7.5	1.305	358	116.44	3354	6.19×10^{-4}	2.1

Keeping in mind Eq. (2.34), the maximum impact force can be obtained as follows:

$$F_{max} = B^{0.4}(2.5\,mgh)^{0.6} \qquad (2.38)$$

The use of the above method is demonstrated with the following example: using a steel ball of 7/4 inch dia. ($R = 22.2$ mm), its mass is 0.357 kg and the drop height is 0.5 m. The potential energy is $mgh = 1.751$ Nm. Measurements were performed on strip flooring elements made of Ash and Spruce. The radii of permanent deformations were 5 and 7.5 mm, respectively. The calculated values are summarised in Table 2.3.

The wood is viscoelastic material over time. Therefore, it is interesting to calculate the average loading velocity which is determined by the deformation z_o and the impact duration T:

$$v = z_0/T \quad \text{and} \quad T = 0.946\frac{\pi}{2}\frac{z_0}{v_0}$$

where v_0 is the falling velocity ($v_0 = \sqrt{2gh} = 3.132 \quad \text{m/s}$).

The impact duration somewhat shorter compared to the linear elastic case (note that the exponent in Eq. (2.34) is 1.5). As Table 2.3 shows, the loading velocity in both cases is 2.1 m/s. This value with all probability is not far from the critical speed corresponding to the infinitely high loading velocity (Sitkei 1994b). Therefore, in this case, the viscoelastic effect should not be taken into consideration.

From certain point of view, the energetic concept might be used also here. The energy transmitted to the body is consumed by permanent deformations which consist of two parts: the visible displaced volume and a compacted volume accommodating the displaced volume (Fig. 2.82). The specific absorption energy is expressed in the following form:

$$E_w = \frac{mgh}{V_d + V_c} = \frac{mgh}{(\psi + 1)V_d} \quad \text{J/cm}^3 \qquad (2.39)$$

where the displaced volume is given by:

$$V_d = \pi z_0^2\left(R - \frac{z_0}{3}\right) \qquad (2.40)$$

and the compressed volume below the surface is approximated as follows:

$$V_c = \psi V_d \quad \text{and} \quad \psi = \frac{\rho_w}{\rho_0 - \rho_w} \tag{2.41}$$

where ρ_0 is the density of compressed wood below the surface, and ρ_w is the volume density of wood.

The compressibility of wood against the infinite half space is an interesting question, and to our best knowledge, it was never investigated. Some observations on the surface compaction of wood due to blunt tool edges, however, have shown that the compacted wood density generally varies between 1.1 and 1.2 g/cm^3. Figure 2.83 shows the expected values of the factor ψ as a function of volume density of wood. Hardwood and softwood show a considerable difference in their compressibility: the Ash has ψ values around 2.2, while the Spruce only around 0.65. Using these values, the total deformed values amount to 0.072 and 0.192 cm^3, respectively, and the specific absorbed energies have values of 24.3 and 9.12 J/cm^3, respectively.

It can be stated that both methods are suitable to characterise wood hardness with acceptable accuracy and resolution. The hardness based on the modulus of elasticity, Eq. (2.36), has better resolution, and it is more suitable to distinguish individual specimens within wood species.

The hardness measurement methods of Janka and Brinell–Mörath are often used also today. They can easily be interpreted in terms of engineering mechanics described above.

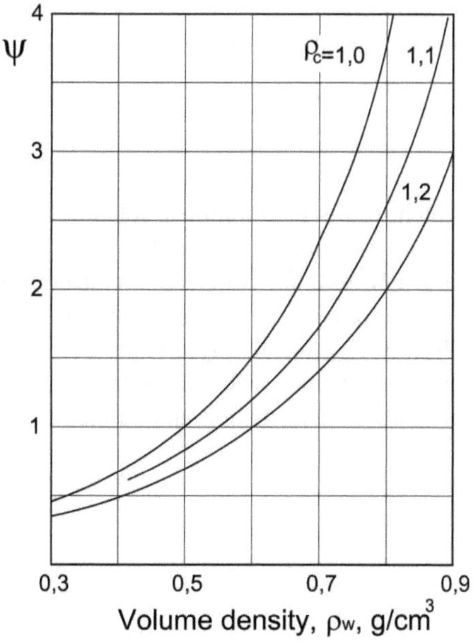

Fig. 2.83 The relative compressibility of wood as a function of initial volume density

Janka (Janka 1912; Janka et al. 1915) used a steel ball of 11.284 mm diameter and pressed half into the wood surface, and the deformed cross section was 100 mm^2. The hardness H_J was defined as the required force F divided by the deformed cross section:

$$H_J = \frac{F}{100} \quad \mathrm{N/mm^2}$$

Substituting the known values into Eq. (2.34), we get

$$\frac{E}{1-\nu^2} = \frac{F}{42.44} \quad \mathrm{N/mm^2}$$

and a comparison of the above equations yields as folows:

$$H_J = 0.4244 \frac{E}{1-\nu^2}$$

giving a linear relationship between the two methods.

The Brinell–Mörath method uses a 10-mm-dia. steel ball and presses it into the wood surface with a given force of 500 N, and the depth of deformation is measured. The force divided by the deformed surface as a spherical calotte gives the hardness, that is

$$H_{BM} = \frac{F}{D\pi z} \quad \mathrm{N/mm^2}$$

where D means the diameter of the deformed surface. Using Eq. (2.34), it can be shown that in this case

$$H_{BM} = 0.5235 \left(\frac{E}{1-\nu^2}\right)^{2/3} \quad \mathrm{N/mm^2}$$

which results in a nonlinear relation between the Brinell–Mörath hardness and the deformation modulus. The following Table 2.4 summarises average hardness values in N/mm^2 for comparison concerning different wood species.

Due to the high portion of plastic deformation at the Janka's method, values of H_J and H_{BM} are hardly comparable. The dynamic method (see Table 2.3) gives higher H_d values, with all probability, due to the elastic recovery which lowers the perceptible deformation and so increases the apparent hardness.

Table 2.4 Comparison of different hardness values

Species	H_J	H_{BM}	$E/(1 - \nu^2)$
Spruce	25	14	140
Pine	33	18	200
Larch	37	20	240
Linden	25	15	155
Beech	60	32	480
Oak	70	34	520
Ash	70	34	520
Hornbeam	75	36	570
Sugar maple	80	40	670
Black locust	100	48	880
Blue gum	80	40	670
Ironwood	130	75	1700
Guaiacum off.	180	86	2100

Note Ironwood = *Olea capensis*, 1020 kg/m^3

2.8 Abrasion Resistance of Wood Surfaces

Utility surfaces of various wood products such as horizontal surfaces of kitchen furniture and flooring in living room are subjected to different, mostly frictional loadings causing wear. The wear process is associated with material loss, deterioration of surface coating, change in colour, loss of gloss and appearance of scratches. All these changes may severely lower the overall utility value of the product, especially its aesthetical value.

The loading conditions of utility surfaces may be quite different, and they are always occasional, concerning a given surface area of probabilistic nature. Therefore, the surface appears to be not evenly worn. Due to the dominating frictional forces, the wear process is mainly abrasive, sometimes enhanced by heat and moisture. Local damages of the surface are generally due to sharp or pointed tools.

The design of utility surfaces requires the careful selection of materials that offer maximum abrasion resistance to all expected loadings. For this reason, the knowledge of abrasion resistance of different wood species is indispensable. If necessary, the abrasion resistance of natural wood surface can be improved using various surface treatments and coatings.

The abrasion resistance of wood surfaces can be determined by the use of different laboratory equipments attempting to imitate real loading conditions. Key factors that must be considered in developing and using an abrasion test include as follows:

- time and kind of contact between the material and the abrading surface,
- the normal load on the abrading surface,
- the speed of the sliding action between the surfaces,

- the temperature at the interface,
- minimum alteration in the surface of abrading material,
- an easy and uniquely defined evaluation of wear and
- a possible correlation between laboratory test results and real wear experiences.

These requirements can only be partially satisfied. Till now, only one standard abrasion procedure is accepted, that is the Taber Abraser. The abraser consists of two rollers coated with sanding paper and mounted on a self-balanced axle. The grade of abrasive material is specified for solid wood, coated and laminated surfaces. The axle is loaded with a given vertical force, and the sample is rotating under the rollers as shown in Fig. 2.84. It is important that the axis of the wheels is shifted from the centre of rotation by about 3/4 inch and the drifting angle is around 30°. The abraded surface is an annulus, and the wear is assessed in terms of quantity of material lost for a given standard condition.

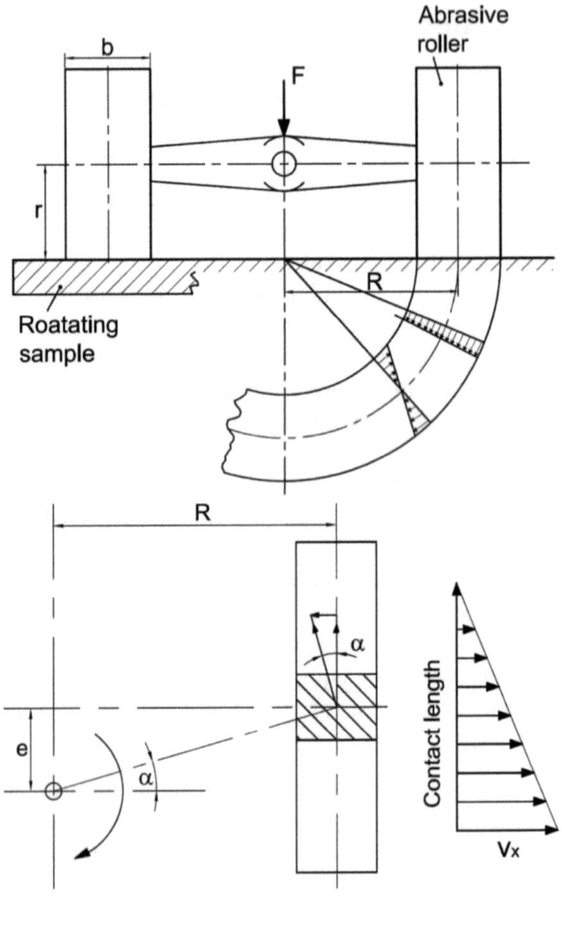

Fig. 2.84 Working principle of Taber Abraser

The sliding action between the material and abrasive rollers is due to the relative motion between them which is characterised by the wheel slip. The resultant slip has three components. The first is due to the rolling resistance of rollers, and the second one is due to the circular motion of the rollers with finite width. The first component is not more than 2–4 %, while the second component is depending on the ratio of the roller width to the diameter of rotation path:

$$S = \pm \frac{b}{2R} \tag{2.42}$$

where
S is the slip,
b is the roller width and
R is the radius of the wheel centre line.

The slip changes its sign at the outer and inner side and has values between 12 and 20 %. It is clear that the slipping of rollers across the width is not constant.

The third slipping component is due to the drifting motion of wheels. The resultant slip in the abraded annular track governs the trajectory of grits during their contact with the wood sample which are depicted in Fig. 2.85. Due to the opposite direction of wheel rotations, the trajectories are crossing each other almost perpendicularly. Using rough sandpaper grade (P-40), the trajectories are fully visible on the abraded surface.

Using the general theory of the sanding process, the working process of the Taber Abraser can be approximated. The quantity of removed material related to the unit width is given by the following equation (Csanády and Magoss 2013):

$$\frac{V\gamma}{b} = BLpv_c t \tag{2.43}$$

where
V is the removed material volume,
γ is volume weight of the wood material,

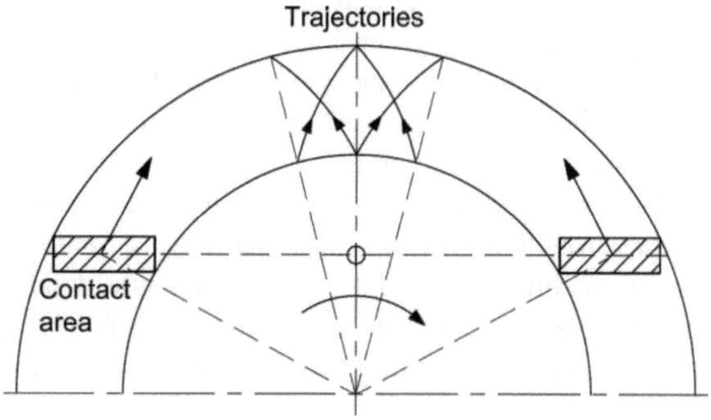

Fig. 2.85 Grit trajectories in the annular track

b is the width of rollers,
B is a material constant related to the unit width, 1/cm,
L is length of sanding path,
p is the surface pressure,
v_c is the sanding velocity and
t is the elapsed time.

The length of the sanding path is equal to the distance done in one rotation, $L = 2R\pi$. An estimate for the sanding velocity requires some considerations. The velocity of rollers is simply given by the relation:

$$v_r = 2R\pi n \quad \mathrm{cm/min}$$

where n is the rotation speed, rpm. The actual sliding velocity is proportional to the slip, and therefore, the sanding velocity is approximated as follows:

$$v_c = 2R\pi nS \tag{2.44}$$

where S is the average slip in the contact surface.

The estimation of the contact pressure may also cause some troubles. The contact pressure for a cylindrical body contacting the plane surface is proportional to the square root of the line load on the wheels. At the same time, the contact width and thus the contact surface is also proportional to the square root of the line load. Therefore, theoretically, the line load on the wheels proportionally increases the removed material. The theory assumes, however, that the rollers are rigid and the surface is perfectly smooth. None of these assumptions are valid in this case.

The rollers are covered with sandpaper, and its deformation behaviour under loading may be very uncertain. The sandpaper by its own deformation increases the contact surface and lowers the pressure in the contact area.

A further fundamental problem is the surface roughness of the wood sample. With rigid wheel and smooth surface, the deformations under the rollers fall into the range of 0.5–1.0 μm, which is much smaller than the roughness heights of the surface.

Examining the Abbott material ratio curves (see Fig. 4.12 in Sect. 4.4), we see that using sharp tools at least to a depth of appr. 20 μm, we may expect layers dense enough to carry the given load. Taking deformation of 20 μm, the corresponding contact width will be appr. 2 mm. Nevertheless, it seems to be well-established that the pressure under the wheels is at least one magnitude higher than those used in the common sanding operation.

It was observed that after a "running-in" period (some 100 revolutions), the surface roughness in the track will have a stationary value corresponding to the given conditions related to sandpaper grade, load on the rollers and the abraded material. In this case, the wear process on the surface will also be stationary and only the continuous wear of the abrasive material decreases the removed material as

Table 2.5 Roughness of the machined Beech sample and the abraded surface

	R_z	R_a	R_q/R_a	R_{ku}	R_{sk}	R_{pk}	R_k	R_{vk}	$R_a/\sqrt{R_k R_z}$
Initial	40.3	5.1	1.37	4.9	−1.1	4.6	12.7	15.7	0.226
	40.6	4.3	1.47	8.5	−1.2	5.9	9.5	9.5	0.218
Track	74.7	11.9	1.25	3.2	−0.4	10	37	23.5	0.226
5 N/cm	76.5	11.2	1.3	3.8	−0.04	14.3	36.5	20.5	0.211
Track	77.3	12	1.3	3.8	−0.5	10.5	40.7	25	0.213
10 N/cm	70	10.9	1.3	3.8	−0.6	10.1	32.5	24.5	0.23

a function of time. The latter can be described with exponential function which makes possible the evaluation of measured results.

As an example, Table 2.5 shows roughness measurements on a machined Beech sample in two directions perpendicular to each other and in the track abraded with 5 and 10 N/cm line load. It is interesting to note that a load change from 5 to 10 N/cm in the same abrading process has made an abnormal wear process in the next 100 revolutions. The difference in roughness shown in Table 2.5, would give no reasons for a deviation in the wear process. A closer look at the grit trajectories in the track, however, revealed that the slightly higher slip at 10 N/cm load will rearrange the trajectories, and the grits recarve the ridges running along the previous trajectories. As a consequence, the amount of the abraded material will be considerably higher than in the stationary process.

As it is already mentioned above, the performance of sandpapers does not remain constant as a function of time. It decreases, with good approximation, according to an exponential function of the form:

$$\frac{\mathrm{d}G}{\mathrm{d}t} = G_0 \mathrm{e}^{-\beta t} \tag{2.45}$$

where G_0 is the initial wear rate and β characterises the decrease in the performance of the sandpaper. The reciprocal value of β is the often used time constant T.

The integral of the above equation gives the removed material as a function of time:

$$G = \frac{G_0}{\beta}\left(1 - \mathrm{e}^{-\beta t}\right) \quad \mathrm{mg/cm} \tag{2.46}$$

Figure 2.86 shows experimental results on Beech samples with two different loads on the rollers as a function of rotation number. The course of the material loss fully obeys Eq. (2.46), except the first point at $N = 100$ rotations. From these results, the time constant β and the material constant G_0 can uniquely be determined. Their values are $\beta = 0.0339$ 1/min (or $T = 29.5$ min) and $G_0 = 37$ mg/cm min.

A stationary wear process was assumed, and the variation in material loss was attributed to the continuous wear of the sand paper. In the first phase of the wear

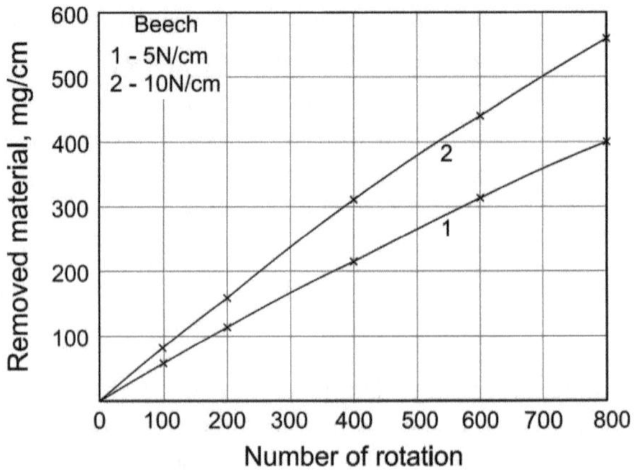

Fig. 2.86 Material loss of Beech samples as a function of number of rotations with two axle loads

process, however, the surface roughness may enhance the wear significantly. In Fig. 2.86, the material loss at N = 100 rotations was 9 % higher than value extrapolated from the overall courses of the wear curve. This deviation is probably due to the influence of the initial roughness on the dynamics of the wear process.

Using rough abrasive materials (P-32, P-40) with grit diameters around 400 μm, in the initial phase each particle can bite a full chip from the relatively smooth surface. As the surface becomes rough with deep furrows, the bite of some grits may be less resulting in a lower wear rate.

In the following, we assume that Eqs. (2.45) and (2.46) are formally valid also for the initial phase of the wear process with the difference that in the first, relatively short phase, the time constant is determined by the change of surface roughness due to the continuous wear. Using the experimental results for Beech wood, the approximate dynamics of the wear process is represented in Fig. 2.87. The two curves with different time constants have the common form:

$$\frac{\mathrm{d}G}{\mathrm{d}t} = G_{0\mathrm{i}}\mathrm{e}^{-\beta_{\mathrm{i}}t} \quad \mathrm{mg/min} \tag{2.47}$$

where G_{0i} is the initial wear rate and β_i is the time constant in the particular section. The elapsed time (t) is given by the number of rotations t = N/60.

During the first 100 rotations, the surface roughness affects the wear rate with a time constant of β_1 = 0.154 1/min (T = 6.5 min) and the initial wear rate is 44.3 mg/min. After 100 rotations, a stationary wear process develops and the further moderate decrease in the wear rate is governed by the deteriorating performance of the abrading material. The stationary wear process would have an initial rate of 37 mg/min with a time constant β_2 = 0.0339 1/min (T = 29.5 min). It is interesting to note that the time constant of the stationary wear process is not far from those found for sanding (T varies between 30 and 60 min).

The two curves in Fig. 2.87 should be determined under the following conditions:

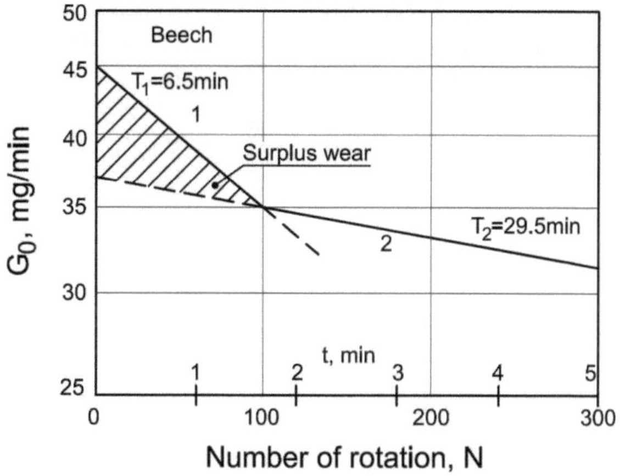

Fig. 2.87 Change of the wear rate in the initial and the stationary phase with different time constants for Beech wood. $P' = 5$ N/cm

- The stationary wear curve (2) is the best fit for all experimental points except the first point at N = 100 rotations.
- The curve for the initial wear phase (1) intersects the stationary curve at N = 100 rotations, and its integral between zero and 100 rotations gives the experimentally obtained value at N = 100 rotations.

The wear rate does not follow the theoretically expected values with different roller loads (0.5 or 1.0 kg). If the load is doubled, the wear rate increased by only some 40 %. There may be many reasons for this. The most important factor is probably the relatively high surface pressure in the contact area. The surface is deeply furrowed at 0.5 kg load, and with a further increased load, the grits cannot proportionally remove more material. Therefore, it is wise to use the same wheel load in comparison studies.

Coated utility surfaces (flooring parquet) were tested for abrasion resistance in the same way with the difference that a finer grade of abrasive sandpaper, P-150, was used. Figure 2.88 shows experimental results for clear coated parquet made of Jatoba (Hymenaea courbaril). To get reliable result for a detailed evaluation, the measurement was continued up to 1000 rotations with 0.5-kg wheel load. The measured points obey the exponential law and allow a reliable evaluation.

During the first 100 rotations, the running-in period, the wear rate considerably exceeds the stationary value (Fig. 2.89). Therefore, it is not recommended to form a judgement based on the results of the first 100 rotations.

The abrasion resistance of wood surfaces can be classified based on measurement results. A possible classification is given in Table 2.6.

The value of "mg/100 rotations" should be taken for the first stationary section of the curve. Generally, the second 100 rotations are suitable for this purpose. The determination of G_0 values is well defined as shown in Figs. 2.87 and 2.89. The abrasion resistance of wood materials may have correlation with their mechanical properties, especially with their hardness. Figure 2.90 represents measurement

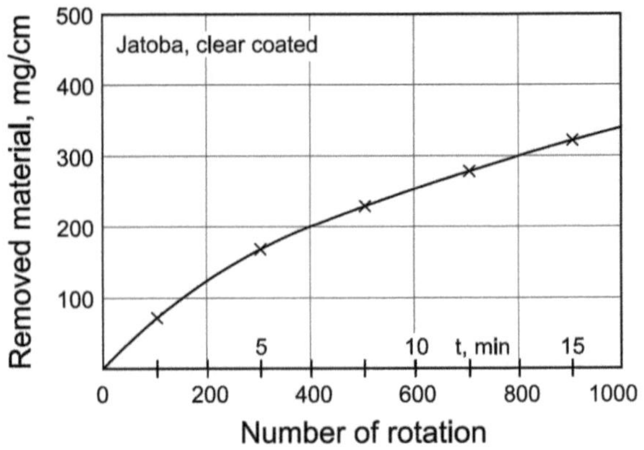

Fig. 2.88 Removed material as a function of the number of rotation on clear coated parquet

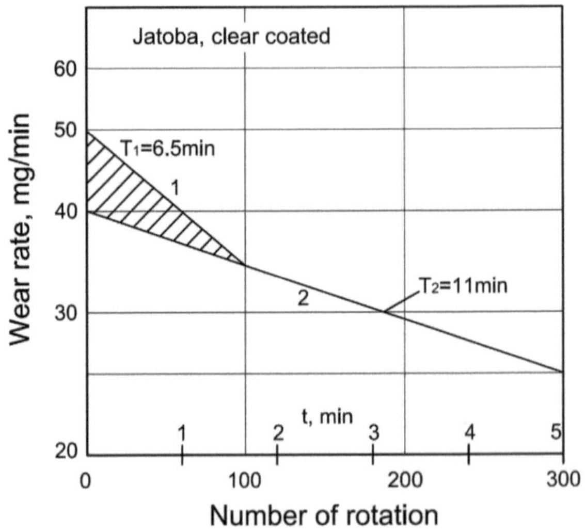

Fig. 2.89 Wear rate as a function of number of rotation for clear coated parquet

Table 2.6 Abrasion resistance classification

mg/100 rotation	G_0 (mg/min)	Classes
<80	<50	Very good
80–150	50–95	Good
150–220	95–140	Satisfactory

results using different wood species. The dynamic hardness was measured with a steel ball of 44.4 mm dia. (7/4 inch) and drop height of 0.5 m. Light conifers have H_d values between 400 and 700 J/cm^3 (or N/mm^2), but dense Pinewood over 700 kg/m^3 had hardness near 2000 J/cm^3.

Black locust has an excellent abrasion resistance at hardness values of 6000–8000 J/cm^3. The very dense Ironwood (1040 kg/m^3) possess hardness values up to 15,000 J/m^3 and correspondingly high abrasion resistance.

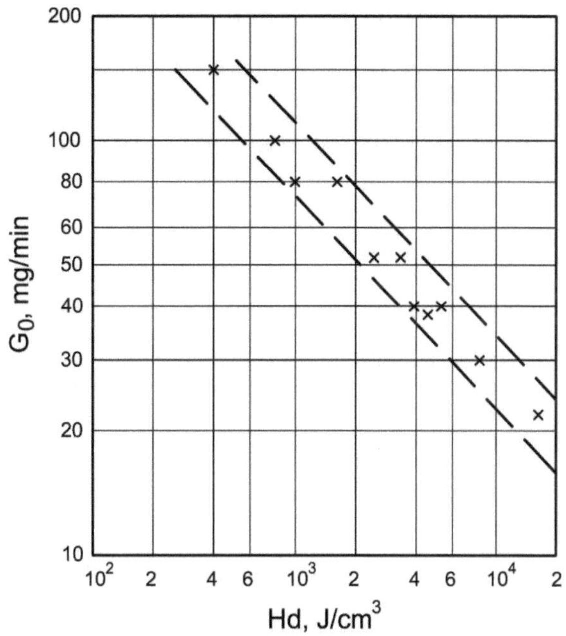

Fig. 2.90 Relation between wear rate and dynamic hardness for different wood species

It should be stressed that the dynamic hardness measurement is not very accurate. The permanent deformation can be measured, but the elastic recovery remains unknown. The size and mass of the ball and the energy of impact may also have a given influence on the obtained result. Furthermore, equation of the contact mechanics used for calculations are only valid, strictly speaking, for small deformations. In this respect, a bigger size is more favourable. Nevertheless, Fig. 2.90 gives a good orientation and overview of abrasion resistance for practical purposes.

References

Agresti, G., Bonifazi, G., Calienno, L., Capobianco, G., Lo Monaco, A., Pelosi, C., Picchio, R., Serranti, S.: Surface investigation of photo-degraded wood by colour monitoring, infrared spectroscopy, and hyperspectral imaging. J. Spectrosc. **1**(1), Article number 380536 (2013)

Ayadi, N., Lejeune, F., Charrier, F., Charrier, B., Merlin, A.: Colour stability of heat-treated wood during artificial weathering. Holz als Roh- und Werkstoff **61**, 221–226 (2003)

Bak, M., Németh, R., Csordós, D., Tolvaj, L.: Effect of treatment medium on the moisture uptake rate and colour change during natural weathering of heat treated wood, pp. 80–86. In: Conference of COST Action FP0904, Rogla, Slovenia, 16–18 Oct 2014

Bak, M., Németh, R., Tolvaj, L.: The colour change of oil-heat-treated timber during weathering. Óbuda Univ. E-Bull. **3**(1), 339–345 (2012)

Béldi, F., Bálint, J.: Einige Sorptionseigenschaften von Zementgebundenen Spanplatten. Acta Facultatis Ligniensis **1**, 25–38 (1986)

Béldi, F, Szabó, J.: Forgécslapok páradiffúziós vizsgálata. Experimental investigation of water-vapour diffusion of particle boards, pp. 43–49. EFE Tudományos Közleményei (1979)
Béldi, F., Bálint, J.: Über einigen Sorptionseigenschaften der Akazie. Acta Facultatis Ligniensis, **1**, 15–27 (1984)
Burmester, A., Olsen, C.: Verbesserung der Formbeständigkeit von Buchenholz durch Tränkung mit Diisocyanat. Holz als Roh- und Werkstoff, S. 84–89 (1971)
Calienno, L., Lo Monaco, A., Pelosi, C., Picchio, R.: Colour and chemical changes on photodegraded beech wood with or without red heartwood. Wood Sci. Technol. **48**, 1167–1180 (2014)
Carslaw, H., Jäger, J.: Conduction of Heat in Solids. Oxford University Press, New York (1959)
Chang, T.C., Chang, H.T., Wu, C.L., Chang, S.T.: Influences of extractives on the photodegradation of wood. Polym. Degrad. Stab. **95**, 516–521 (2010)
Christensen, G., Kelsey, K.: Die Geschwindigkeit der Wasserdampfsorption durch Holz. Holz als Roh- und Werkstoff, S.178–188 (1959)
Cloutier, A., Fortin, Y.: A model of moisture movement in wood. Wood Sci. Technol. **27**, 95–114 (1993)
Cloutier, A., Fortin, Y.: Moisture content—water potential relationship from saturated to dry conditions. Wood Sci. Technol. **25**, 263–280 (1991)
Csanády, E., Magoss, E.: Mechanics of Wood Machining. Springer, Berlin (2013)
Fowkes, F.M.: Acid-base interactions in polymer adhesion. In: Mittal, K.L. (ed.) Physicochemical Aspects of Polymer Surfaces, vol. 2, pp. 583–603. Plenum Press, NY (1983)
Gardner, D.J.: Application of the Lifshitz—van der Waals acid-base approach to determine wood surface tension components. Wood Fiber Sci. **28**(4), 422–428 (1996)
Gardner, D.J., Generella, N.C., Gunells, D.W., Wolcott, M.P.: Dynamic wettability of wood. Langmuir **7**, 2498–2502 (1991)
Garrett, H.: Contact angles and their significance for adhesion, pp. 19–41. In: Proceedings of Northampton College of Advanced Technology (1964)
Gindl, M., Sin, G., Gindl, W., Reiterer, A., Tschegg, S.: A comparison of different methods to calculate the surface free energy of wood using contact angle measurement. Colloids Surf. A **181**, 279–287 (2001)
Good, R.J.: Contact angle, wetting, and adhesion: a critical review. In: Mittal, K.L. (ed.) Contact Angle, Wettability and Adhesion, pp. 3–36. VSP, Utrecht (1993)
Hill, C.A.S.: Wood Modification: Chemical, Thermal and Other Processes. Wiley, London (2006)
Huang, A., Zhou, Q., Liu, J., Fei, B., Sun, S.: Distinction of three wood species by Fourier transform infrared spectroscopy and two-dimensional correlation IR spectroscopy. J. Mol. Struct. **883–884**, 160–166 (2008)
Huang, X., Kocaefe, D., Kocaefe, Y., Boluk, Y., Pichette, A.: Study of the degradation behavior of heat-treated jack pine (*Pinus banksiana*) under artificial sunlight irradiation. Polym. Degrad. Stab. **97**, 1197–1214 (2012)
Huang, X., Kocaefe, D., Kocaefe, Y., Boluk, Y., Krause, C.: Structural analysis of heat-treated birch (*Betula papyrifera*) surface during artificial weathering. Appl. Surf. Sci. **264**, 117–127 (2013)
Janka, G.: Härteprüfung des Holzes mittels Kugeldruckverfahren. Wien (1912)
Janka, G.: Die Härte der Hölzer, Wien (1915)
Kataoka, Y., Kiguchi, M., Fujiwara, T., Evans, P.D.: The effect of within-species and between-species variation in wood density on the photodegradation depth profiles of sugi (*Cryptomeria japonica*) and hinoki (*Chamaecyparis obtusa*). J. Wood Sci. **51**, 531–536 (2005)
Kataoka, Y., Kiguchi, M., Williams, R.S., Evans, P.D.: Violet light causes photodegradation of wood beyond the zone affected by ultraviolet radiation. Holzforschung **61**, 23–27 (2007)
Kubovsky, I., Kacík, F.: Colour and chemical changes of the lime wood surface due to CO_2 laser thermal modification. Appl. Surf. Sci. **321**, 261–267 (2014)
Liptáková, E., Kúdela, J.: Analysis of the wood-wetting process. Holzforschung **48**, 139–144 (1994)

Masuda, M.: Why human loves wood grain figure? Extraction of vision-physical characteristics deeply related to impression, pp. 11–23. In: ICWSF Conference, Ljubljana, 5–7 Sept 2001

Matsuo, M., Umemura, K., Kawai, S.: Kinetic analysis of color changes in keyaki (*Zelkova serrata*) and sugi (*Cryptomeria japonica*) wood during heat treatment. J. Wood Sci. **60**, 12–20 (2014)

Matsuo, M., Yokoyama, M., Umemura, K., Gril, J., Yano, K., Kawai, S.: Color changes in wood during heating: kinetic analysis by applying a time-temperature superposition method. Appl. Phys. A **99**, 47–52 (2010)

Mitsui, K.: Changes in the properties of light-irradiated wood with heat treatment. Part 2. Effect of light-irradiation time and wavelength. Holz als Roh- und Werkstoff **62**, 23–30 (2004)

Mitsui, K., Tsuchikawa, S.: Low atmospheric temperature dependence on photodegradation of wood. J. Photochem. Photobiol. B **81**, 84–88 (2005)

Mitsui, K., Takada, H., Sugiyama, M., Hasegawa, R.: Changes in the properties of light-irradiated wood with heat treatment. Part 1. Effect of treatment conditions on the change in color. Holzforschung **55**, 601–605 (2001)

Mitsui, K., Murata, A., Tolvaj, L.: Change in the properties of light-irradiated wood with heat treatment. Part 3. Monitoring by DRIFT spectroscopy. Holz als Roh- und Werkstoff **62**, 164–168 (2004)

Mitsui, K., Tolvaj, L., Papp, G., Bohus, J., Szatmári, S., Berkesi, O.: Changes in the properties of light-irradiated wood with heat treatment. Part 4. Application of laser. Wood Res. Slovakia **50**, 1–8 (2005)

Müller, U., Rätzsch, M., Schwanninger, M., Steiner, M., Zöbl, H.: Yellowing and IR-changes of spruce wood as result of UV-irradiation. J. Photochem. Photobiol. B **69**, 97–105 (2003)

Nakshabandi, G., Kohnke, H.: Thermal conductivity and diffusivity of soils as related to moisture tension. Agr. Meteorol. **2**, 271–279 (1965)

Németh, K.: Rolle des Feuchtigkeitsgehaltes des Holzes in der Hygroskopizität von Polyuretan-Holz-Kompositen. Acta Facultatis Ligniensis Sopron, S. 55–64 (1986)

Pandey, K.K.: Study of the effect of photo-irradiation on the surface chemistry of wood. Polym. Degrad. Stab. **90**, 9–20 (2005a)

Pandey, K.K.: A note on the influence of extractives on the photo-discoloration and photo-degradation of wood. Polym. Degrad. Stab. **87**, 375–379 (2005b)

Pandey, K.K., Vourinen, T.: Comparative study of photodegradation of wood by a UV laser and a xenon light source. Polym. Degrad. Stab. **93**, 2138–2146 (2008)

Persze, L., Tolvaj, L.: Photodegradation of wood at elevated temperature: colour change. J. Photochem. Photobiol. B: Biol **108**, 44–47 (2012)

Popescu, C.M., Popescu, M.C., Vasile, C.: Structural analysis of photodegraded lime wood by means of FT-IR and 2D IR correlation spectroscopy. Int. J. Biol. Macromol. **48**(4), 667–675 (2011)

Popescu, M.-C., Froidevaux, J., Navi, P., Popescu, C.-M.: Structural modification of *Tilia cordata* wood during heat treatment investigated by FT-IR and 2D IR correlation spectroscopy. J. Mol. Struct. **1033**, 176–186 (2013)

Sell, J.: Physikalische Vorgänge in wetterbeanspruchten Holzbauteilen. Holz als Roh- und Werkstoff, S. 259–267 (1985)

Sitkei, G. (ed.).: Faipari műveletek elmélete (Theory of wood processing), pp. 105–141. Szaktudás Kiadó Budapest (1994a)

Sitkei, G.: Non-linear rheological method describing compaction processes. Int. Agrophys. **8**, 137–142 (1994b)

Srinivas, K., Pandey, K.: Photodegradation of thermally modified wood. J. Photochem. Photobiol. B **117**, 140–145 (2012)

Stamm, A.: Passage of liquids, vapours and dissolved materials through softwoods. USDA Bull. **929**, 1946

Tolvaj, L., Faix, O.: Artificial ageing of wood monitored by DRIFT spectroscopy and CIE L*a*b* color measurements. I. Effect of UV light. Holzforschung **49**, 397–404 (1995)

Tolvaj, L., Mitsui, K.: Surface preparation and direction dependence of DRIFT spectra of wood. Appl. Spectrosc. **58**, 1137–1140 (2004)

Tolvaj, L., Mitsui, K.: Light source dependence of the photodegradation of wood. J. Wood Sci **51**, 468–473 (2005)
Tolvaj, L., Mitsui, K.: Validity limits of Kubelka-Munk theory for DRIFT spectra of photodegraded solid wood. Wood Sci. Technol. **45**, 135–146 (2011)
Tolvaj, L., Varga, D.: Photodegradation of timber of three hardwood species caused by different light sources. Acta Silvatica et Lignaria Hungarica **8**, 145–155 (2012)
Tolvaj, L., Molnar, S., Nemeth, R., Varga, D.: Color modification of black locust depending on the steaming parameters. Wood Res. Slov **55**, 81–88 (2010)
Tolvaj, L., Z, Molnar, Nemeth, R.: Photodegradation of wood at elevated temperature: infrared spectroscopic study. J. Photochem. Photobiol. B **121**, 32–36 (2013)
Tolvaj, L., Molnar, Zs., Magoss, E.: Measurement of photodegradation-caused roughness of wood using a new optical method. J. Photochem. Photobiol., B **134**, 23–26 (2014a)
Tolvaj, L., Nemeth, R., Pasztory, Z., Bejo, L., Takats, P.: Colour stability of thermally modified wood during short-term photodegradation. Bioresources **9**(4), 6644–6651 (2014b)
Tolvaj, L., Popescu, C.M., Molnar, Zs., Preklet, E.: Dependence of the air relative humidity and temperature on the photodegradation processes of beech and spruce wood species. Bioresources (in print) (2015)
van Genuchten, M.: A closed-form equation for predicting the hydraulic conductivity of unsaturated soils. Soil Sci. Soc. Am. J. **44**, 892–898 (1980)
Van Oss, C.J., Chaudhury, M.K., Good, R.J.: Monopolar surfaces. Adv. Colloid Interface Sci. **28**, 35–64 (1987)
Varga, D., van der See, M.E.: Influence of steaming on selected wood properties of four hardwood species. Holz als Roh- ond Werkstoff **66**, 11–18 (2008)
Tolvaj, L., Tsuchikawa, S., Inagaki, T., Varga, D.: Temperature dependence of photodegradation of wood monitored by colour measurement. Wood Sci. Technol. doi:10.1007/s00226-015-0749-1 (in print)
Yildiz, S., Yildiz, U.C., Tomak, E.D.: The effects of natural weathering on the properties of heat-treated alder wood. Bioresources **6**(3), 2504–2521 (2011)
Zavarin, E., Jones, S.J., Cool, L.G.: Analysis of solid wood surfaces by diffuse reflectance infrared Fourier transform (DRIFT) spectroscopy. J. Wood Chem. Technol. **10**, 495–513 (1990)
Лыков, А.В., Михайлов, Ю.: А. Теория переноса энергии и вещесва Изд. АН. БССР, Минск, 1959. (Theory of energy and mass transfer)
Шубин, Г.: Физические основы и расчет процессов сушки древесины, pp. 271–279. Лесная промышленность, Москва (Physical background of wood drying) (1965)

Chapter 3
Colour Characterisation of Wood

3.1 Introduction

One of the most characteristic features of human eye is to see and distinguish colour. The origin of colours is the electromagnetic radiation in the visible region with different reflection distributions. Depending on the reflectance distribution, our eye perceives a given colour by its lightness and hue.

The colour of various timbers varies in a very wide range from the almost white through yellow, orange, brown and red to pitch-black. Due to the continuously rising character of the reflectance distribution as a function of wavelength, the colour of wood is always mixed.

The colour of wood is a very important aesthetical property, and it has a fundamental influence on the commercial value of timbers. Nevertheless, the use of colour measurements and evaluations in practice is unsatisfactory also today. In order to make the use of colour measurement and evaluation more widespread also in the all-day design practice, a new colour estimation method has been introduced. Furthermore, new colour interpretations, representations and practical applications are presented. Some examples of the new multi-angle measurement are also demonstrated. Scientifically, well-established explanations are given for colour changes due to irradiation, different heat treatments and weathering. A quite new topic is the gloss of wood surfaces which is receiving ever more attention today. Up to now no comprehensive study on gloss properties came to light in the literature. Based on detailed experiments, the most important relations of the gloss of wood surfaces are presented with many practical applications.

E. Csanády et al., *Quality of Machined Wood Surfaces*,
DOI 10.1007/978-3-319-22419-0_3

3.2 The Colour of Wood

The most important aesthetical characteristics of wood materials are their colour including the delicate differences in colour hues (inhomogeneity of colour) and their pattern and glossiness. The most precious timbers have always distinct colour hues, beautiful figures, pattern and texture giving a natural image to our eyes. It is suggested that correlations should be existing between our psychological images and vision-physical characteristics of wood (Masuda 2001). Psychological images of our eyes can not exactly be measured and, therefore, their correlation with the measurable visible physical properties is always conditional depending on the observer. Nevertheless, most of the people agree that natural wood always gives us warm and natural images.

Visible light is a small portion of the electromagnetic spectrum. The wavelength of the visible light ranges from 360 to 760 nm (Fig. 3.1). Ultraviolet (UV) light is at the lower end of this range and infrared (IR) light is at the upper end. Physically, these rays are not coloured. Colour is only recognised by the human eyes when rays enter the eye and they are absorbed by light-sensitive receptor cells. The sensitivity of human eyes to the visible light depends and varies with wavelength. Since this sensitivity cannot directly be measured, this causes some uncertainty in all colour estimation method.

In general, materials absorb a part of a given wavelength and reflect the rest. The reflected light will be recognised by the human eyes as a colour which is dependent on the spectral distribution of the reflected light. If we consider selective wavelength in the visible range, then different basic colours can be distinguished (Fig. 3.1). For example, the blue colour is found in the range from 400 to 500 nm and the red colour around 700 nm. The spectral reflectance curves of different wood species are virtually always continuous, and therefore, we have a mixed colour. That means, for example, that we have radiation and reflection in the blue range (400–500 nm), but we never see a wood with blue colour.

The spectral reflectance curve in the visible region accurately and uniquely represents the colour of a wood sample. In order to represent the colour numerically, measurement and evaluation methods have been developed. In general, these are based on the determination of trichromatic qualities, which means that any colour can be made by mixing three basic colours (i.e. red-green-blue).

The human vision is approximated by colour parameters obtained from measurements with instruments. Generally, a spectrophotometer is used giving the

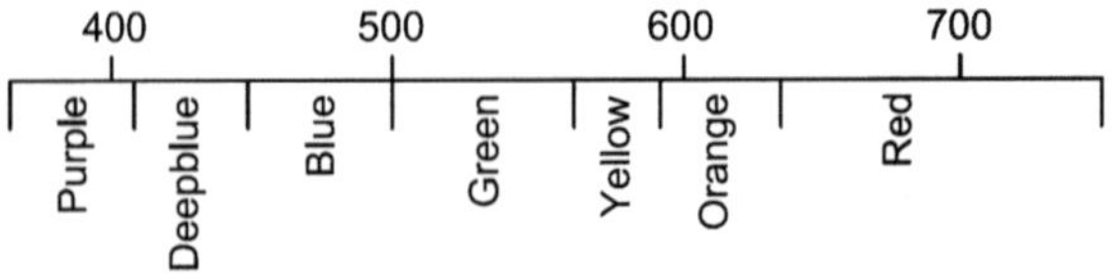

Fig. 3.1 The visible range of the electromagnetic radiation and its colour hues

reflectance distribution curve with readings in 10-nm steps. At the same time, the measurement unit calculates all standardised colour parameters.

The most frequently used colour system is the CIE (Comission Internationale de l'Eclairage) $L^*a^*b^*$ colour space. Using the CIE visibility function $V(\lambda)$, the X, Y and Z tristimulus values are integrated and from these the following colour characteristics will be calculated (Wyszecki and Stiles 1969):

$$L^* = 116\left(\frac{Y}{Y_0}\right)^{1/3} - 16$$

$$a^* = 500\left[\left(\frac{X}{X_0}\right)^{1/3} - \left(\frac{Y}{Y_0}\right)^{1/3}\right]$$

$$b^* = 200\left[\left(\frac{Y}{Y_0}\right)^{1/3} - \left(\frac{Z}{Z_0}\right)^{1/3}\right]$$

where

L^* is the characteristic of lightness or darkness, and its value is between zero and 100

a^* is the positive coordinate, for redness and negative for greenness

b^* is the positive coordinate for yellowness and negative for blueness

The calculations have to be done for definite standard colour temperature values of the illumination. This three-dimensional colour space is schematically demonstrated in Descartes coordinates in Fig. 3.2. The lightness L^* is determined alone by the tristimulus value Y and calculated mainly from the reflection between 500 and 600 nm. Concerning the existing wood colours, a^* and b^* are always positive. The b^*/a^* ratio is generally higher than one, but for deep-red colours there are values less than one.

The drawback in plotting rectangular coordinates is that the most important colour characteristics, the colour hue, is not explicitly shown. This drawback,

Fig. 3.2 The CIE $L^*a^*b^*$ colour space

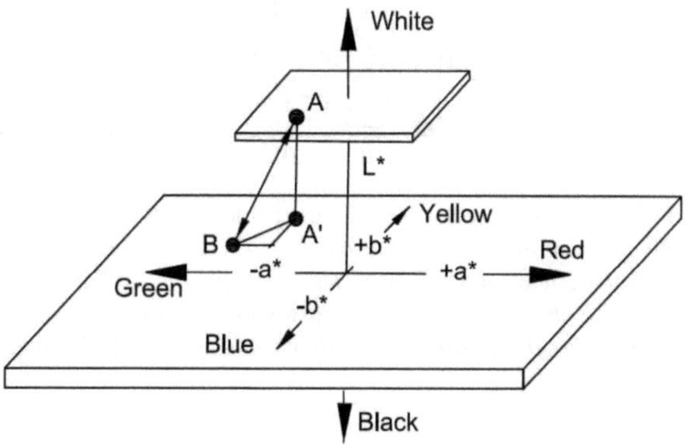

however, can easily be eliminated drawing in the lines of constant b^*/a^* ratios which define the colour hue (see in Fig. 3.3 and later in Sect. 3.6). Polar coordinates can also be used to transform the rectangular coordinates into polar ones. The radius will be given as follows:

$$C = \sqrt{a^{*2} + b^{*2}}$$

which is termed *colour saturation*, and the angle is calculated as follows:

$$h^* = \arctan\left(\frac{b^*}{a^*}\right)$$

which is the colour hue. It is clear that the *colour hue* interpreted in rectangular or polar coordinates is uniquely the same although the numerical values are different. In the following, we generally use the $L - a^* - b^*$ system and, therefore, the $h^* = b^*/a^*$ definition.

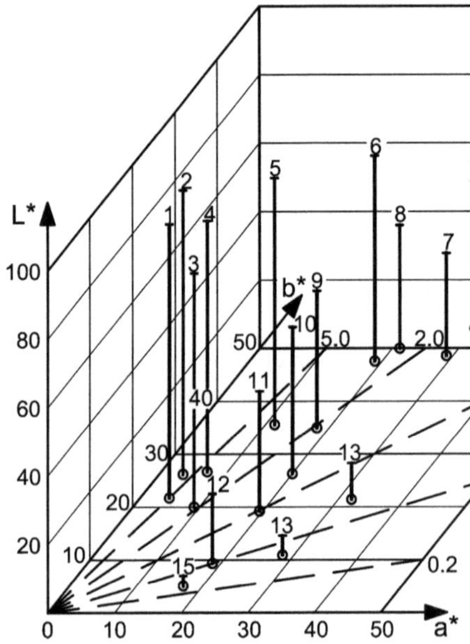

Fig. 3.3 Wood species with different colour hues on the $b^* - a^*$ plane. (*1*) *Fraxinus excelsior*, (*2*) *Tilia tomentosa*, (*3*) *Fagus sylvatica*, (*4*) *Pinus sylvestris*, (*5*) *Lagarostrobos franklinii*, (*6*) *Pinus sylvestris*, latewood, (*7*) *Caesalpinia sappan*, (*8*) *Xanthocyparis vietnamensis*, (*9*) *Dacrydium cupressinum*, (*10*) *Fitzroya cupressoides*, (*11*) *Eucalyptus diversicolor*, (*12*) *Acacia donaldsonii*, (*13*) *Dalbergia maritima*, (*14*) *Berchemia zeyheri*, (*15*) *Xanthorrhoea preissii*

Examining the effect of different surface treatments, the determination of colour change may be useful in the following form (see Fig. 3.2):

$$\Delta E^* = \left[(\Delta a^*)^2 + (\Delta b^*)^2 + (\Delta L^*)^2\right]^{1/2}$$

It should be stressed that this numerical value does not give any information about the real change in lightness or colour hue! Latter can only be characterised by the ΔL^* and Δh^* values.

As an example, Table 3.1 shows the colour characteristics for different wood species. Here, it should also be mentioned that the CIELAB system does not give any correspondence between colour characteristics and real colour hues! Therefore, seeing the colour characteristics in Table 3.1, one has no idea about the real colour of wood species.

In order to give some imagination on placing of various wood species in the $b^* - a^*$ plane, Fig. 3.3 shows the representation of selected species in a three-dimensional coordinate system also showing their lightness. It can clearly be seen that species with greater lightness occur at high b^*/a^* ratios having greyish colour hues. In this selection, the Red ivory (Berchemia zeyheri) has the highest red colour coordinate a^*. The red-coloured species are placed below the $b^*/a^* = 1$ line. Species with deepest red colours are the Bois de Rose and Black boy.

Concerning light coloured wood species, a particular attention should be paid to the greyish colour. Already Stefan (in 1879) defined the true grey body as having the same reflection independently of wavelength. Indeed, heavily weathered surfaces show a nearly constant reflectance as a function of wavelength (see later in Fig. 3.84).

As the colour is changing towards grey, the reflection is continuously growing in the range from 400 to 500 nm. On the contrary, a vivid reddish colour has very low reflections in this range.

The emissivity and reflectivity of real materials also depend on their surface properties. A surface element of black body radiates in all directions and this radiation is termed as *diffuse radiation*. In this case, Lambert's cosine law holds and the directional emissivity or reflectivity is independent of the angle of viewing. The other extreme case is when the incident light is perfectly mirrored under the same angle as it strikes the surface, Fig. 3.4.

Actually, all materials exhibit a certain degree of dependence of the emissivity or reflectivity on the viewing angle. Due to this directional reflectivity, the *brightness* (or gloss) from a radiating surface that the human eye detects depends on the radiative flux originating from a unit area of the radiating body as it appears to the eye. The apparent brightness is proportional to the directional emissivity or reflectivity. Therefore, a radiating black surface has the same brightness regardless of the viewing angle. Wood and paper materials have the same brightness in a wide range of viewing angles, but over 65°–70° the reflectivity quickly decreases (Schmidt and Eckert 1935; Eckert 1936).

Table 3.1 Colour characteristics of selected wood species

Wood species	*X*	*Y*	*Z*	*a**	*b**	*L**	*h**
Ash	55.99	57.27	40.2	4.27	21.94	80.33	78.98
Linden	59.52	59.07	42.31	3.80	25.81	83.33	81.62
Cotton wood	55.0	54.59	39.86	4.42	23.96	78.64	79.54
Beech	44.58	42.31	33.12	9.09	19.28	71.08	64.75
Oak	39.86	37.85	25.49	8.69	24.74	67.91	70.64
Walnut	24.01	22.6	17.5	8.23	16.03	54.66	62.82
Larch	45.58	41.93	27.48	13.07	26.72	70.82	63.93
Pine	51.73	52.24	32.1	5.88	27.35	77.43	77.87
Pine (grain)	30.72	27.14	7.4	19.58	47.53	59.13	67.62
Tasm. Myrtle	21.16	18.48	8.47	18.45	28.15	50.08	56.76
Huon Pine	43.64	42.09	19.7	11.36	36.15	70.93	72.56
Celery Top P.	40.43	39.11	19.8	10.69	32.48	68.83	71.79
Jarrah	7.63	6.22	1.94	17.76	26.78	29.96	56.45
Karri	10.9	8.25	2.25	25.46	31.91	34.49	51.42
Alerce	20.24	17.32	5.85	20.09	35.66	48.66	60.6
Totara	9.95	7.68	1.38	23.37	38.18	33.3	58.53
Kauri	26.89	24.52	9.02	15.55	37.56	56.6	67.51
Rimu	13.39	11.33	3.05	18.42	35.76	40.14	62.75
Puriri	10.78	10.02	3.32	9.96	30.11	37.88	71.69
Rewarewa	24.21	22.39	10.52	13.6	29.22	54.44	65.05
Tawa	32.53	31.32	15.84	10.49	30.11	62.77	70.79
Widdringtonia	40.13	39.35	25.41	9.0	22.82	69.01	68.48
Stink-wood	18.42	17.0	4.67	12.61	40.42	48.26	72.68
Yellow-wood	25.8	23.54	8.23	15.26	38.52	55.62	68.39
Imbuya	6.74	5.92	1.59	12.31	28.85	29.21	66.9
Olea	11.14	9.36	1.66	17.82	40.96	36.69	66.49
Sugi 1	34.82	33.01	19.83	12.51	24.3	64.17	62.76
Sugi 2	15.2	13.29	6.95	16.46	21.75	43.2	52.88
Hinoki	49.5	48.41	25.42	10.1	33.29	75.08	73.26
Giang huong	10.98	7.97	1.48	28.5	38.08	33.92	53.19
Po-mu	30.08	28.43	11.4	12.24	36.79	60.27	71.6
Bách vang	11.79	9.48	1.1	21.55	49.61	36.9	66.52
Sucupira	6.9	5.95	1.68	13.58	28.08	29.3	64.19
Amgelim pedra	18.6	16.43	7.78	16.7	26.13	47.53	57.41
Bois de Rose	3.01	1.59	0.01	32.64	22.53	13.16	34.62

Note Giáng huong = *Pterocarpus macrocarpus*, Po-mu = *Fokienia hodginsii*, Bách váng = *Xanthocyparis vietnamensis*, Bois de Rose = *Dalbergia maritima*

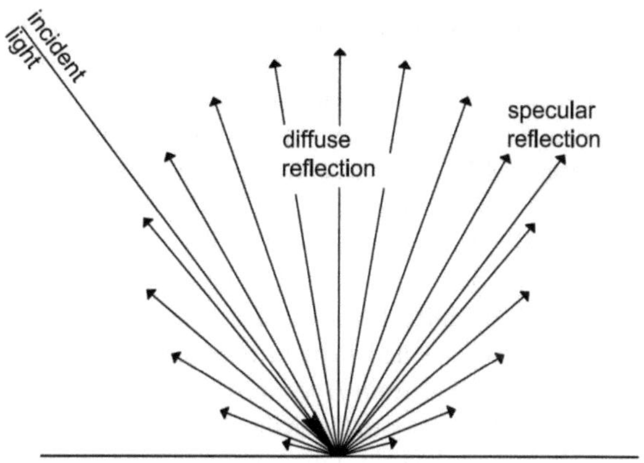

Fig. 3.4 Diffuse and directional radiation of a surface

The local reflectivity of machined wood surfaces, due to their microrelief, can be perceptible different from place to place which may give an additional beauty to the surface. Furthermore, different consumers may prefer different glossiness for their furniture and other wood products. As a consequence, not only the colour but also the gloss of the surface will become ever more important. A more detailed analysis of brightness of wood surfaces will be given later.

3.3 The Origin of Wood Colour

The colour characteristics of wood species depend on the chemical components of the wood structure, formation of various organic substances in the heartwood and on the surface microtopography due to machining or surface treatments. The reaction of wood components to light, heat and other chemicals under the influence of moisture can change the colour of wood. As a "living material", wood is generally prone to discolour with age also in indoor conditions. In outdoor conditions, many different factors participate in the discolouration process and light irradiation, heat, cold and moisture are the major influencing factors.

The main structural substances of wood, cellulose and hemicellulose do not absorb visible light. The third main component, lignin, absorbs and reflects light below 500-nm wavelength and has a pale yellow colour. In general, this yellow colour is typical for the sapwood of almost all wood species.

As a sapwood changes into heartwood, its cavities of latter will be packed with extractives including terpenes, polyphenols, wood resins, tannins, sugars and fatty acids. They make significant contributions to such properties of wood as colour, odour and decay resistance. Some of these materials are reflecting more light at higher wavelength and have deeper colours.

The colouring materials in wood are mainly aromatic systems which are generally more chemically stable than long-chain-conjugated compounds. The basic colours of these compounds are red, yellow and brown. Other colours result from mixtures of these basic colours.

The red colouring agents are mainly the substituted derivatives of naphthoquinone and anthraquinone with varying numbers of different groups in different positions (Mills and White 2011). The characteristic groups are the methyl ($-CH_3$), hydroxyl (–OH), carboxyl (–COOH), carbonyl (=CO) and hydroxymethylene ($-CH_2OH$). The colouring material is a mixture of several components such as alizarin, purpurin, rubiadin, brazilein, haematin, etc. Sappan wood (*Caesalpinia sappan*), also called Brazilwood in trade, has the main constituent of brazilin which is oxidised by air and light to produce brazilein. Haematin is the main colouring agent of logwood from *Haematoxylon campechianum* in Central America.

The yellow-coloured constituents are mostly flavonoid components such as quercitin (as a glucoside) in many wood species, rhamnetin from Rhamnus species, fisetin from *Cotinus coggygria* and *Rhus verniciflua* (Urushi), berberin from Berberis and Phellodendron species. In many cases, several of compounds may exist together giving different shades. For instance, the Brazil wood contains not only brazilein but also quercetin and emodin, producing its unique orange-red colour (Badami et al. 2004). Brown colours are produced by tannins (e.g. *Castanea sativa*) which are polyphenols. Their basic constituents are gallic acid ($C_6H_3(OH)_3COOH$) and/or digallic and trigallic acids (Mills and White 2011).

The age of a tree in many cases influences the quantity and colour of extractive materials in the heartwood considerably. The Karri wood (*Eucalyptus diversicolor*) has darker red colour with age. It is believed, furthermore, that the richness of colour comes also from the quality of soil it grows in. For example, the deepest red Myrtle (*Nothofagus cunninghamii*) comes from highly fertile soils on bazalt.

It is believed, for example, that the characteristic yellow colour of Urushi (*Rhus verniciflua*) is due to flavonoids such as fisetin and sulfetin mainly.

If the tree is infected with a staining fungus, it produces blackheart wood with distinctive dark brown, black and even green streaks running through the wood (Blackheart Sassafras—*Atherosprerma moschatum*). Sassafras timber is renowned in furniture use as a solid wood or veneer, as well as for decorative work, bowl turning and so on.

The surface of wood can modify the colour considerably. The slender wood cells are arranged are layers in one direction. During woodworking operations, these cells are cut in a given plane in relation to the grain. Depending on the direction of cut, a different microphysics-relief, topography can be achieved. Furthermore, the cut of wood surface is always associated with brittle fracturing of cell walls causing diffuse reflectance. Because the microrelief of the surface has generally a definite direction, therefore, the irradiating direction of light may slightly modify the wood colour. Figure 3.5 shows a schematic diagram of light reflection depending on the irradiating direction (Masuda 2001). If the vessels and lumens are cut, bottom surface of lumens reflects the incident light similarly to bundles of small mirrors. The bunch of mirrors are aligned and divided by cells walls. The incident light is much more reflected parallel to the grain (directional reflection) giving a high gloss. At the same time, the light will be more scattered (diffuse reflection) perpendicular

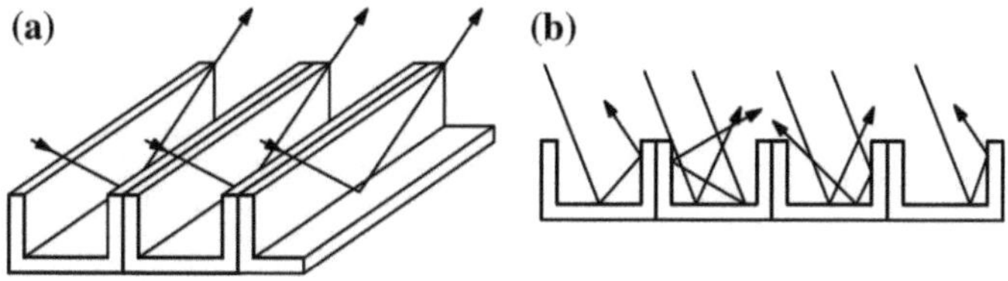

Fig. 3.5 Schematic diagram of reflection of light on wood surface, parallel to the grain (**a**) and perpendicular to the grain (**b**)

to the grain and from the edge of cell walls. For example, the sanding of a wood surface always endows with a greyish blear (veil) which can be removed using wax, or oil treatment. The directional reflection parallel to the grain can considerably modified by breaking the continuity of reflecting mirrors (see in Fig. 3.5) using a proper machining process, for instance, sanding perpendicular to the grains (experimental evidences are given later in Figs. 3.25 and 3.31).

It is interesting to note that change in a given wood surface, depending on its microtopography, may have lesser change in its colour hue than in its glossiness. Gloss is a directional reflection parameter which may be changed depending on many influencing factors such as incident and viewing angles, position of cutting surface in relation to the grain and surface roughness (see in Sect. 3.7).

Some correlation may be found between lightness and colour hue. Concerning quite different wood species, the correlation is very loose, Fig. 3.6.

A much better correlation exists within a particular wood species as demonstrated in Fig. 3.7 for three selected wood species. The St. John's bread has a wide range of colour hues and in spite of the fact, the relationship strongly follows a given asymptotic curve. Generally, all curves have an asymptotic shape, but the initial tangent of the curves may vary considerably. From these species, the Sugi has the highest scattering especially in the brownish red range.

Fig. 3.6 Relationship between lightness and colour hue for different wood species

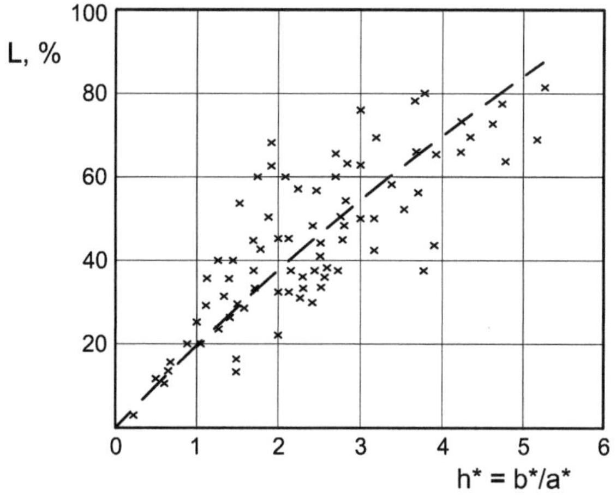

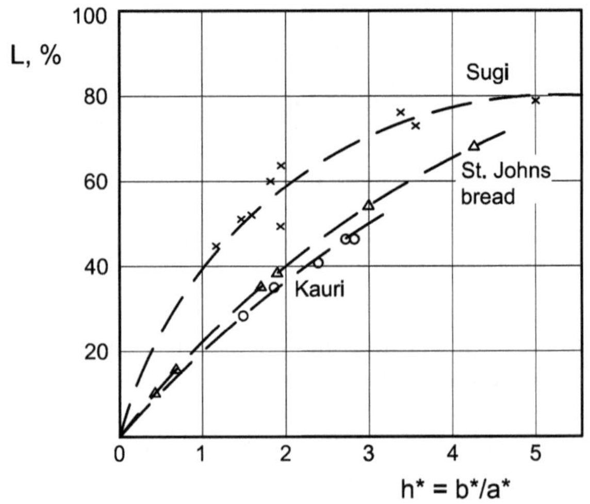

Fig. 3.7 Lightness versus colour hue for Sugi (*Cryptomeria japonica*), Kauri (*Agathis australis*) and St. John's bread (*Ceratonia siliqua*)

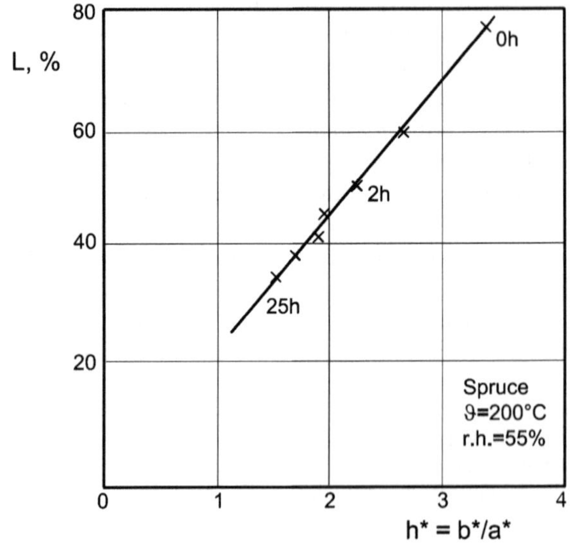

Fig. 3.8 Lightness and colour change of Spruce during heat treatment. Air temperature 200 °C, relative humidity 55 %

The $L^* - h^*$ relationship is generally observed as linear after heat treatment on a given wood species. This explained the fact that the range of colour change is smaller and, therefore, the measurement points can be approximated with a straight line. As an example, the relationship between lightness and colour hue for Spruce is plotted in Fig. 3.8 using high temperature heat treatment (Bekhta and Niemz 2003). The change is highest in the first two hours when the initial yellowish colour turns into light brown and then into brownish red in 25 h.

3.4 A New Colour Estimation Method

From the aesthetic point of view of the colour of wood materials, our eyes appreciate a given surface based on its lightness, colour hue and the inhomogeneity of colour (delicate differences in various pattern). Therefore, in all colour description and evaluation, methods require a uniquely defined characterisation of colour regarding its three main characteristics mentioned above. Furthermore, based on a measurement record (spectral reflectance curve), the observer should give numerical values for an entire colour characterisation and also, since only our eyes recognise colour, a correspondence principle between numerical values and colour hues. This ensures that someone can imagine a colour in full details which was given numerically. The recent standardised methods do not fully meet the above requirements.

Several other questions can be put concerning standardised methods. In the CIELAB system, the lightness is determined only by the green tristimulus value (Y) according to a third power equation. It is hard to believe that other parts of the reflection spectrum do not contribute to the perceptible lightness. In many experimental works, the evaluation contains a^* and b^* values separately which alone do not characterise a colour. In fact, values a^* and b^* may show the same tendency of increase or decrease and this means that the colour hue does not change. The definition of colour hue in the rectangular $L - a^* - b^*$ system is missing although it could be used in the simple $h^* = b^*/a^*$ form.

Due to the above shortcomings of existing methods, many valuable experimental results have a low power of expression, especially for the practice. Therefore, further improvements in the colour representations are needed for better understanding the colour phenomenon and to make the existing experimental results more usable in practice.

The starting point in all colour determination is the measurement of the reflection spectrum in the visible region which most accurately represents the colour of a material. The observation of a vast amount of spectral reflectance curves for very different wood species has shown that the different curves (Fig. 3.9) can be described with acceptable accuracy in the following general form (Sitkei 2013):

$$R = a + b \cdot \lambda^n \tag{3.1}$$

where R is the reflectance, λ is the wavelength and a, b and n are the constants characterising a given wood species.

It is interesting to mention here that a surface whose absorptivity or reflectivity is the same for all wavelengths is called a grey surface (Stefan 1879). This makes the exponent n in Eq. (3.1) is equal to zero. Concerning the existing different wood species, the exponent n varies in a wide range between 0.4 and 4.0.

The following general rule can be observed about the spectral reflectance curve of different wood species given in Fig. 3.9 and their colour hue. Starting from a true grey colour, the reflection is continuously decreasing mostly in the range of 400–500 nm

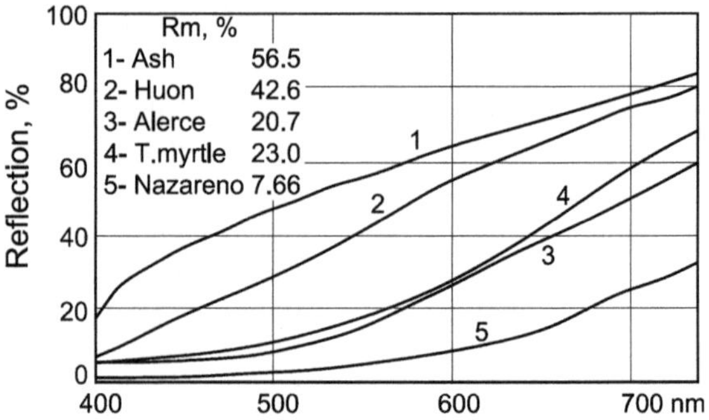

Fig. 3.9 Reflection spectra for different wood species

and the colour is also changing from grey yellow to yellow and further to yellow, yellow-brown, brown, brown-red and finally to red. At the same time, the exponent n varies from 0.5 (grey-yellow) to 3.0 (brown-red).

The use of Eq. (3.1) allows a convenient mathematical treatment for revealing some basic relationships between the shape of the reflection curve and the colour hue belonging to it.

The area under the reflection curves defines the true lightness termed as *physical lightness*. The physical lightness R_m is given by integration of the reflection curve in the visible region from wavelengths of 400–700 nm.

$$R_m = \left(a \cdot \lambda + \frac{b}{n+1} \lambda^{n+1} \right) \Big/ 3 \tag{3.2}$$

It is worth to put the origin of the coordinate system to the wavelength of 400 nm and to take $x = \lambda/100$.

Due to the specific shape of the curves, the average reflection R_m, as a horizontal line, intersects the reflection curve at a wavelength λ_0 uniquely characterising the spectral distribution of reflectance (Fig. 3.10) and also the colour hue.

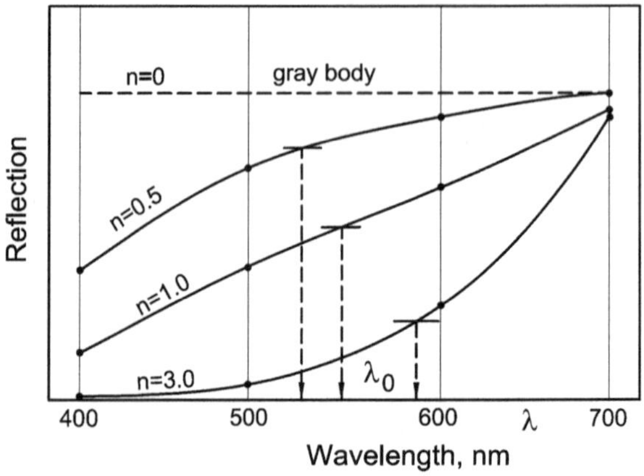

Fig. 3.10 Intersection points (λ_0) for different spectral distributions

The point of intersection λ_0 can be calculated as follows:

$$\lambda_0 = \left(\frac{R_m - a}{b}\right)^{1/n} \cdot 100 + 400 \tag{3.3}$$

The intersection point does not depend on the initial intersection (a) of the reflection curve at 400 nm and, therefore, a simpler equation can be derived (Sitkei 2013):

$$\lambda_0 = \frac{300}{(n+1)^{1/n}} + 400 \tag{3.3a}$$

Equation (3.3a) shows that the value of λ_0 is uniquely defined by the exponent n and the relationship is plotted in Fig. 3.11. Analysing wood samples of different colour, a correspondence could have been established between λ_0 values and colour hues given also in Fig. 3.11.

The next problem is now the determination of the exponent n from the spectral reflectance curves for each wood species. Using the reflection data, the curves can be fitted according to Eq. (3.1) using standard procedures. However, the plot of reflection curves on double logarithmic scale has revealed that the exponent n cannot always be considered as a constant. Furthermore, to perform the approximation of very large data system is a time-consuming procedure.

In order to avoid this problem, we sought out an other characteristic number uniquely related to the exponent n. One possible choice was the following number composed of distinct ordinates of the reflection spectrum:

$$K_1 = \frac{R_{550} - R_{450}}{R_{650} - R_{550}} \tag{3.4}$$

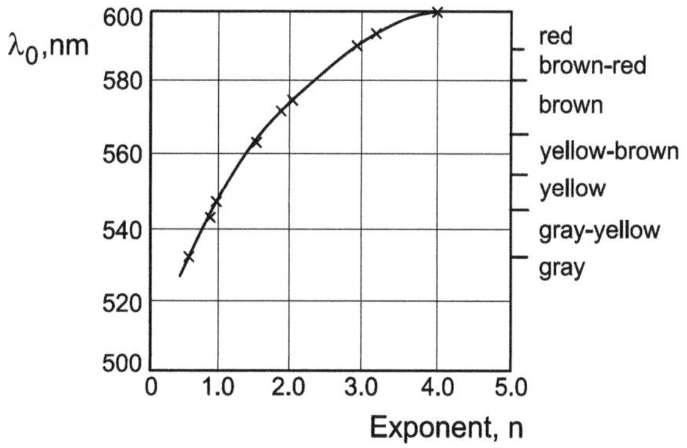

Fig. 3.11 Characteristic wavelength λ_0 as a function of exponent n and the corresponding colour hues

where the subscripts refer to the given wavelength. This number is independent of the initial section of the reflection curve (a) because the following equality holds:

$$\frac{(R_{550}-a)-(R_{450}-a)}{(R_{650}-a)-(R_{550}-a)}=\frac{R_{550}-R_{450}}{R_{650}-R_{550}}$$

Therefore, the characteristic number K_1 is the only function of the exponent n. Using Eq. (3.1) and taking $a = 0$ and $x = \lambda/100$, Eq. (3.4) can be expressed in the following form:

$$K_1=\frac{1.5^n-0.5^n}{2.5^n-1.5^n} \tag{3.4.a}$$

Both K and λ_0 are the functions of the exponent n only; therefore, a uniquely defined relationship between K_1 and λ_0 can be established which is given in Fig. 3.12. The proof of usability of this relationship will be given later using and plotting the corresponding measurement results in the same coordinate system.

The characteristic number K can also be calculated in a somewhat different manner. The ordinates R_{450}, R_{550} and R_{650} are the central ones in the three sections of the reflection spectrum. Concerning a linear spectral distribution, the ordinate R_{450} is numerically equal to the average reflection between 400 and 500 nm, that is, the integrated area under the reflection curve in this section (R_{m1}). In other cases, when $n \neq 1$, the central ordinates do not accurately coincide with the average reflection values. Therefore, an other possibility is to express the characteristic number K with the corresponding average reflections. The average reflections in the three sections can be calculated quite similar to Eq. (3.2), integrating Eq. (3.1)

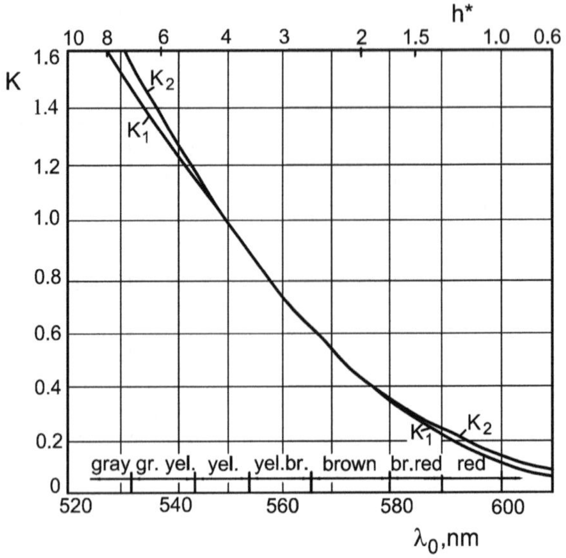

Fig. 3.12 Theoretical relationship between K and λ_0 with colour hue ranges

between the pertinent upper and lower limits. Replacing the ordinates R_{450}, R_{550} and R_{650} with the corresponding average reflections R_{m1}, R_{m2} and R_{m3}, the following theoretical relationship is obtained

$$K_2 = \frac{2^{n+1} - 2}{3^{n+1} - 2^{n+2} + 1} \tag{3.5}$$

which is plotted also in Fig. 3.12. The course of the curve in the central part is just the same as for K_1. There are some deviations in the upper part of the curve compared to K_1. The comparison of this theoretical relationship with experimentally obtained values will be given in the following (see Fig. 3.13).

In practice, a spectrophotometer measures and supplies the reflection ordinates for every 10 nm in the visible region. In the following, we use the range from 400 to 700 nm which contains a total of 31 data. Numerical integrations between the appropriate upper and lower limits give the average partial reflections R_{m1}, R_{m2}, R_{m3} and R_m for the entire range. A simple numerical procedure uses the subsequent ordinates, for example

$$R_{m1} = \frac{R_{400} + R_{410} + R_{500}}{11}$$

A slightly better accuracy can be achieved by using the average values of two neighbouring ordinates, for example:

$$R_{m1} = \frac{(R_{400} + R_{410})/2 + \cdots + (R_{490} + R_{500})/2}{10}$$

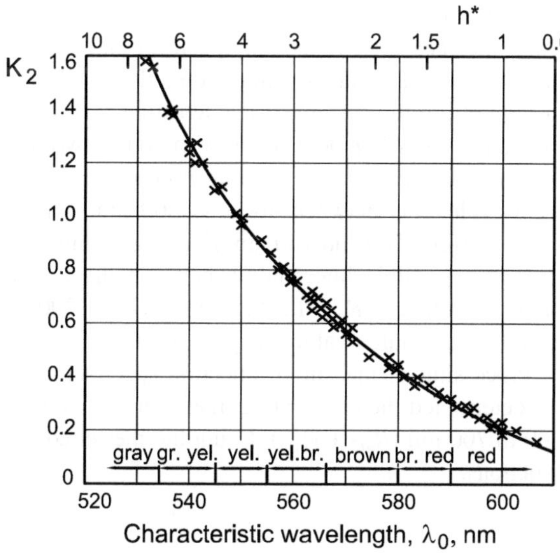

Fig. 3.13 Plot of experimental results in a $K - \lambda_0$ coordinate system

In order to determine the characteristic numbers K_1 or K_2 based on experimental results, the following equations will be used:

$$K_1 = \frac{R_{550} - R_{450}}{R_{650} - R_{550}} \quad \text{and} \quad K_2 = \frac{R_{m2} - R_{m1}}{R_{m3} - R_{m2}}$$

The experimental value of λ_0 will be given by finding the intersection point between the horizontal line of R_m and the spectral curve (see Fig. 3.10).

Knowing the experimentally obtained K_1 and λ_0 or K_2 and λ_0 coherent values, they will be plotted in a coordinate system similar to Fig. 3.12. Because there is an agreement between the theoretically derived curve and the experimentally obtained data points, this method is useful for a quick colour estimation. In the case of a good agreement namely, the K_1 or K_2 value alone determines the expected value of λ_0 and also the colour hue!

The results based on the measured spectral reflectance curves for several different wood species are presented on Fig. 3.13. The measured points follow the theoretical curve with acceptable accuracy, and the relative scattering is low enough to determine the colour hue with absolute certainty and accuracy. The horizontal axis below contains also the range of colour hues for quick estimation. The resolution of colour coordinates enables to differentiate within a colour range. On the upper side of the diagram, the corresponding standardised colour scale (h) is also given for comparison and control. The correspondence of λ_0 to h^* will be discussed later.

3.5 Interrelations of the CIELAB System

In the CIELAB system, the X, Y and Z tristimulus values will be determined by using the measured reflection spectrum and the sensitivity distribution curves of human eyes. The latter cannot directly be measured, and this produces some uncertainty in the system. It is interesting to note that the blue sensitivity curve of the human eyes is independent of the other two, and the green and red sensitivity curves, however, are overlapping each other. The blue sensitivity curve occupies the wavelengths between 400 and 500 nm, while the green and red occupy the range between 500 and 700 nm. Keeping in mind the above statements, it is evident to seek a possible interrelation between the Z tristimulus values and the partial physical lightness R_{m1}. In the range of 500–700 nm, however, a combination of X and Y tristimulus values may be used.

Processing of measured reflection spectra for some 30 different wood species, we determined the physical lightness for the blue range R_{m1} and for the range from 500 to 700 nm (R_{m2} + R_{m3}). Using the measured X, Y and Z tristimulus values, the functions.

$Z = f(R_{m1})$ and $\sqrt{X \cdot Y} = f(R_{m2} + R_{m3})$ are plotted on logarithmic scale, Fig. 3.14. The figure shows an interesting picture. The function $Z = f(R_{m1})$ is strongly linear with a small shift which can be expressed in the following form:

$$R_{m1} = \frac{z}{1.1} \tag{3.6}$$

The function $\sqrt{X \cdot Y} = f(R_{m2} + R_{m3})$ has an exponent of 1.28 and its equation is written as follows:

$$\sqrt{X \cdot Y} = 0.276(R_{m2} + R_{m3})^{1.28}$$

or

$$(R_{m2} + R_{m3}) = 2.73 \cdot (X \cdot Y)^{0.39} \tag{3.7}$$

The sum of the above two equations gives the total physical lightness

$$R_m = \frac{1}{3}\left[\frac{Z}{1.1} + 5.47(X \cdot Y)^{0.39}\right] \tag{3.8}$$

Due to the small scattering range of experimental data in Fig. 3.14, Eq. (3.8) has a good accuracy.

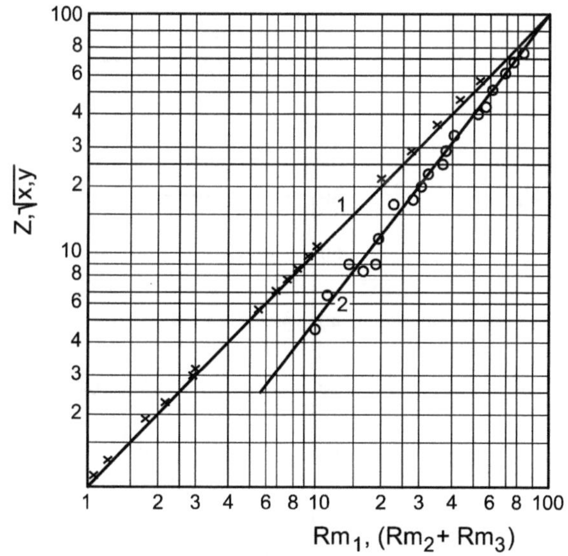

Fig. 3.14 Interrelations between the colour coordinates and partial physical lightness. $1 - Z = f(R_{m1}), 2 - \sqrt{X \cdot Y} = f(R_{m2} + R_{m3})$

Figure 3.15 shows a relationship between the physical lightness R_m and the standardised lightness L. Here, we have a wider scattering zone due to the fact that the lightness L is calculated from only the green tristimulus values which are physically not correct. The relationship is approximated with the equation

$$R_m = 0.0264 \cdot L^{1.75}\,\% \tag{3.8a}$$

If we assume that the major part of the reflection occurs in the 500–700 nm range, then a relationship between the colour tristimulus values $\sqrt{X \cdot Y}$ and the physical lightness R_m can be found (Fig. 3.16). The correlation is fairly strong which is described as follows:

$$R_m = 2.68(X \cdot Y)^{0.385} \tag{3.8b}$$

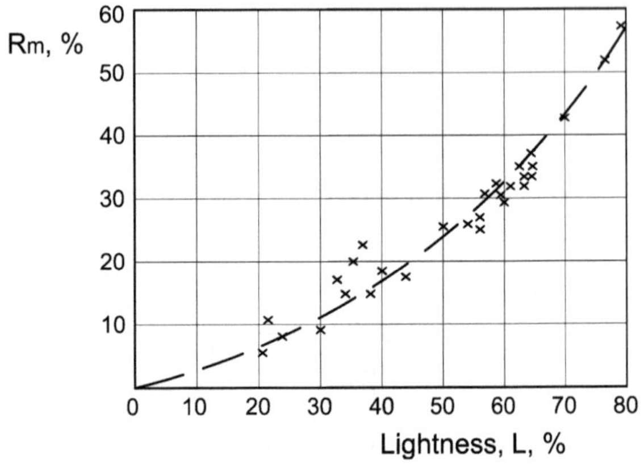

Fig. 3.15 Physical lightness R_m as a function of standardised lightness L

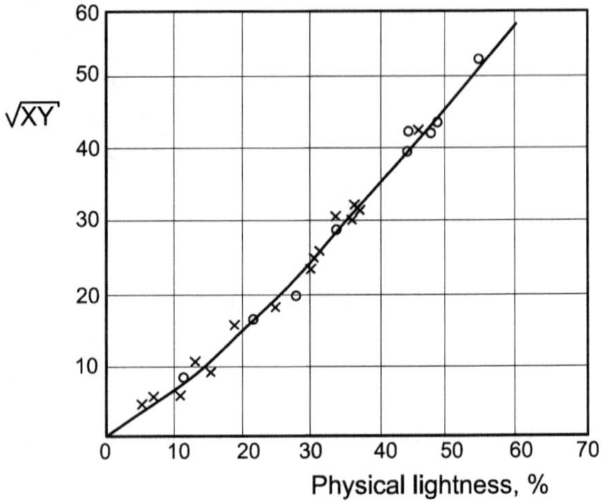

Fig. 3.16 Relationship between physical lightness and X, Y coordinates

The exponent in Eq. (3.8b) is quite similar to that of Eq. (3.7).

An other interesting finding that the physical lightness of the middle zone R_{m2} is equal to the green tristimulus values Y with surprisingly high accuracy, that is $R_{m2} = Y$. At the same time, R_{m3} has a weak correlation with the red tristimulus values X (Sitkei 2013),

Finally, an important relationship between the two colour hue characteristics, λ_0 and h^*, should be examined. In this study, h^* was calculated as the ratio of chromaticity indices b^* and a^*. The correlation between λ_0 and h^* is given in Fig. 3.17, which is rather week, especially for smaller values of h^*. This means that the use of h^* for colour hue determination may be uncertain, especially for woods with brownish red and red colours. The resolution of the λ_0 scale seems to be much better for deeper colours.

A correlation equation between λ_0 and h^* can be approximated in the following simple form

$$\lambda_0 = \frac{600}{(h^*)^{0.062}}$$

Based on the above theoretical considerations and experimental results, the following main conclusions may be drawn:

- Concerning wood materials, the reflection spectrum alone is suitable to determine their physical lightness and colour hue,
- the characteristic wavelength λ_0 determines the colour hue with considerable accuracy and its resolution is better than that of the standardised parameter h^*,
- the characteristic number K is suitable alone to estimate the expected colour hue, and
- there are interesting interrelations among the standardised colour characteristics and the introduced new parameters. The physical lightness can be determined from the X, Y and Z tristimulus values.

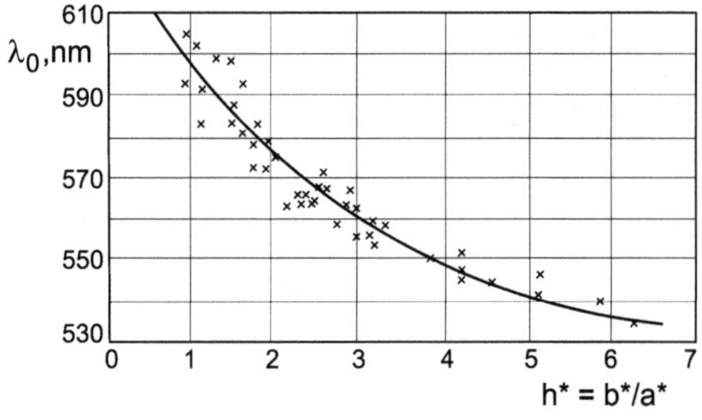

Fig. 3.17 Interrelation between the colour hue characteristics λ_0 and h^*

3.6 Multi-angle Colour Measurement

Reflection measurements on technical surfaces in the past already revealed that the angle of illumination considerably influences reflection (Schmidt and Eckert 1935; Eckert 1936). Later investigations on wood samples confirmed that the wood colour, especially the lightness, may be slightly different according to the irradiating direction of the light (Minemura and Umehara 1979). Using an incidence angle of 45°, the lightness varied with some 10 % as the irradiation direction changed in comparison with the fibre direction. The highest lightness was measured when the incident light was perpendicular to fibre direction. Values of a^* and b^* had a small change in the same phase, and therefore, the colour hue remained almost constant.

New measurements with a multi-angle spectrophotometer gave similar results. Although the conventional and multi-angle units have different measuring principles, the obtained results are more or less comparable with each other. The used multi-angle unit has a 45° illumination and six viewing angles which are related to the illumination angle as shown in Fig. 3.18. Using an Ash sample (*Fraxinus excelsior*), the measurement results along and perpendicular to the grains are depicted in Fig. 3.19.

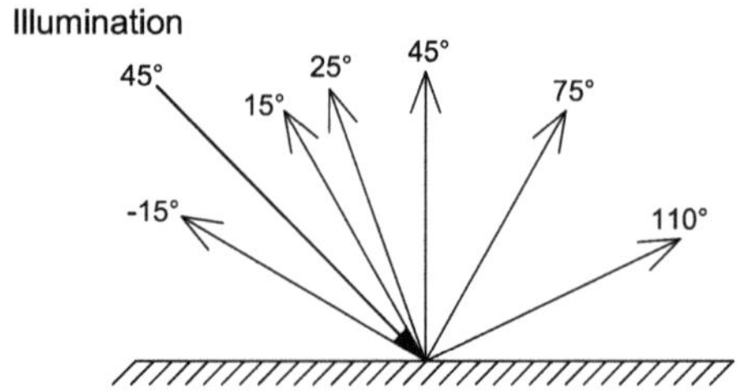

Fig. 3.18 Angle of illumination and six viewing angles of a multi-angle spectrophotometer

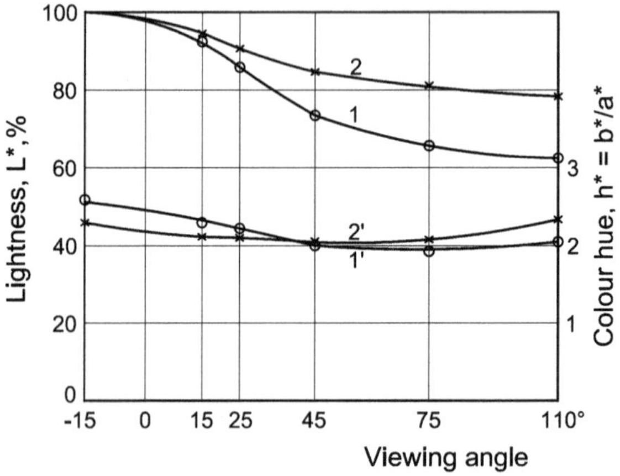

Fig. 3.19 Multi-angle measurement on a planed Ash sample (*Fraxinus excelsior*). (*1,1′*) Along the grains, (*2,2′*) perpendicular to the grains. Standard values: L_0^* = 80.6 %, h^* = 4.85, L_{max}^*/L_0^* = 1.24

Starting from −15° viewing angle, the lightness continuously decreases. The decrease along the grains is roughly twice as much compared to that of viewing angle perpendicular to the grains. The colour hue first is also slightly decreasing but at higher viewing angles it remains constant or very little increases.

Altogether it can be stated that the change in colour hue is generally unessential. At the same time, the higher lightness at viewing across the grains considerably contributes to the *grain enhancement* on the given surface. This effect cannot be observed using conventional measurement techniques.

As a further example, Fig. 3.20 shows measurement results on a swamp Kauri sample (*Agathis australis*, 40,000 years old). The relative change in lightness is considerable and much higher than in the previous case. This may simply be attributed to the lower lightness of the Kauri wood. For the naked eyes, the Kauri sample does not seem to be as bright as it would be from the measurement at −15° viewing angle.

In certain cases, the wood structure may irregularly influence the lightness variation as a function of viewing angle. Figure 3.21 shows an example for Balsa wood compared with two other species. Balsa is one of the lightest woods among the existing timber species with high void ratio. As a consequence, its surface rather is a diffuse reflector decreasing the dependence of its lightness on the viewing angle. It is striking, furthermore, that also the change of its colour hue is almost independent of the viewing angle. All other species tested uniformly show higher h^* values at viewing angle of −15° compared to their average colour hue, but Balsa is an exception.

Keeping in mind the above examples, it may be stated that lightness of a surface is more sensitive to the viewing angle, quite similarly to gloss variations. The colour hue is a much more constant material property and less prone to variations.

For comparison purposes, Fig. 3.22 represents the reflection spectrum with multi-angle measurement and the standard reflection spectrum. The course of the

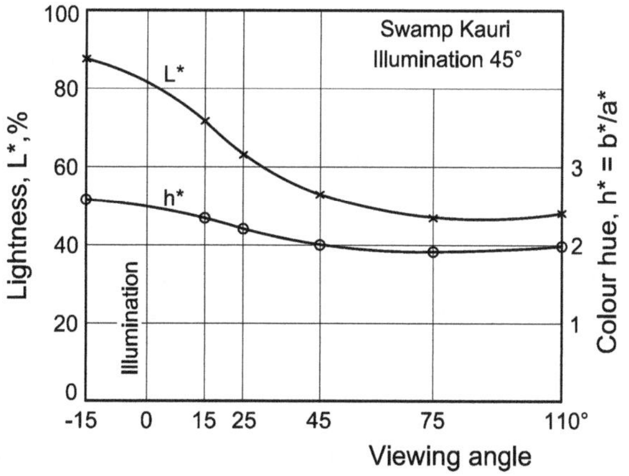

Fig. 3.20 Lightness and colour hue variation as a function of viewing angle related to the illumination angle 45°. Measurement along the grains. Standard values: $L_0^* = 56.6$ %, $h^* = 2.37$, $L_{max}^*/L_0^* = 1.54$

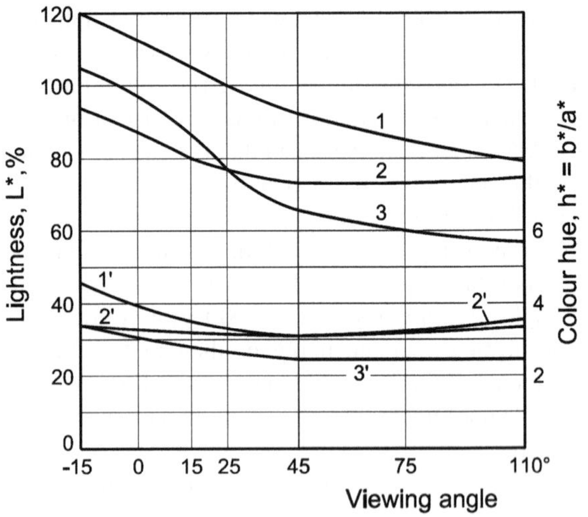

Fig. 3.21 Effect of viewing angles on lightness and colour hue. (*1,1'*) Beech (*Fagus sylvatica*), (*2,2'*) Balsa (*Ochroma pyramidalis*), (*3,3'*) Mahagony (*Swietenia mahagoni*)

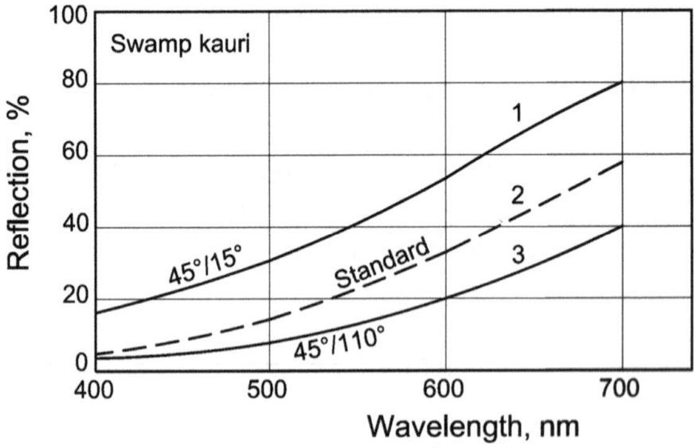

Fig. 3.22 Comparison of multi-angle reflection spectra and standard spectrum for Kauri wood. (*1*) $K = 0.694$, $h^* = 2.4$, $\lambda_0 = 562$ nm, (*2*) $K = 0.62$, $h^* = 2.37$, $\lambda_0 = 566$ nm, (*3*) $K = 0.5025$, $h^* = 1.97$, $\lambda_0 = 572$ nm

reflection spectra is similar but those measured with extreme viewing angles are considerably shifted in relation to the standard reflection spectrum. It is interesting that the standard reflection spectrum approximately corresponds to +30° to 35° viewing angle with 45° illumination.

Finally, the multi-angle measurement technique creates new possibilities first of all in the evaluation of lightness differences along and across the grains.

3.7 Gloss of Colour Surfaces

The most important aesthetical properties of a wood surface are its colour, glossiness and the visible structural figures including inhomogeneities. The gloss of surface is the results of light reflections due to the incident light from different directions. The reflection from a wood surface is very complicated process due to the distinct and highly anisotropic texture of wood. As a consequence of the anisotropy, a new plane surface cut from a bole can be done in arbitrary positions in relation to the main axis of bole, and therefore, in principle infinite number of different structure pattern of a surface could be achieved. If we take only the most common length, radial and tangential cuts, the possible variations in texture profiles and figures are even considerable.

Another major influencing factor is the surface roughness. Due to the annual growing of a tree, the size of vessel and tracheid diameters in the earlywood and latewood are considerably different. As a consequence, for the same machining condition the local roughness and its reflection on the earlywood and latewood will also be different. Furthermore, if the direction of cut is not parallel with the fibres, diving fibres occur with irregular reflection properties. Irregular growth in wood materials (burl, curly wood, streaked, bird's eye and fiddle-back) will also cause different reflection properties giving a gentle inhomogeneity to the surface.

The angle of incident light has also a definite influence on the reflection properties of a surface. This influence is highly dependent on the reflection ability of the given surface requiring us to rack our brains in the selection of appropriate incident and measuring angle. In order to avoid such difficulties, standardised measuring angles, generally 20°, 60° and 85°, are established and recommended, but this does not fully eliminate these problems.

In particular, the gloss of wood surfaces has not been measured and analysed yet. As outlined above, wood surfaces, due to their distinct structure, may have quite different reflection behaviour. For example, the principal portion of the structure of soft woods consists of longitudinal tracheids and much less of ray parenchyma tissues. Furthermore, the diameters of early and latewood cells differ considerably. The hardwood species have different principal cell and tissue types including vessels, fibres and rays in earlywood and latewood. The diameter of vessels may have values up to 300 μm and their distribution within the trunk may be diffuse-porous and ring-porous. Cutting the wood in a longitudinal direction, the vessels and tracheids cut in half may act as small mirrors considerably enhancing the local reflection. Wood materials characteristically have a brittle fracture and they, on the other hand, will reflect much rather similarly to a diffuse reflector. Therefore, the treatment of machined wood surfaces (sanding, oil, wax) may change the reflection ability substantially.

3.7.1 Measurement of Gloss

As outlined above, reflection properties of wood surfaces are highly dependent on many factors such as the angle of incident light, surface roughness, the cut cell structure on the surface, direction of grains and the viewing angle. One extreme case is the fully diffuse reflection which means a constant reflection in all directions. The other extreme case is the specular reflection (perfect mirror) where reflection occurs only under the mirrored angle. Neither of these extreme cases occurs in natural circumstances as outlined already in Sect. 3.4. (see Fig. 3.3).

The emissivity and reflectivity of real surfaces fundamentally depend on their surface properties such as surface roughness, the density of wood (void ratio), the character of brittle fracture of cells remaining on the surface and the illumination angle in relation to the grains. Due to their microrelief, the local reflectivity of machined wood surfaces can change from place to place giving an additional beauty to the surface. The gloss of a wood surface should generally depend on the direction of viewing or measurement (Eckert 1936).

The sanding of a wood surface always endows with a greyish blear (veil) which can be removed using oil or wax treatment. It is interesting to note that a given wood surface, depending on its microtopography, may have change lesser in its colour hue than in its glossiness. Gloss is a directional reflection parameter, and therefore, it may be changed depending on many influencing factors such as incident and viewing angles, positions of cutting surface in relation to the grain and surface roughness (microtopography).

Due to the need, especially in the automobile and lacquer industries, but in the last time also in furniture and parquet flooring industries, brightness (glossiness) measurement methods were developed and standardised. A light beam which strikes a surface under a given incident angle partly will penetrate the layer of paint or surface of the material and the rest will be reflected. For surfaces of different reflectability, an optimum incident and viewing angle was established and recommended. The more common measuring angle is the 60° but angles of 20° for high-gloss and 85° for low-gloss surfaces are recommended. In general, the most common 60° measuring angle is suitable for wood surfaces.

A gloss metre does not indicate the percentage of reflected light, which may cause some trouble with interpretation of readings using different incident angles. A black-polished glass standard is used with a given refractive index (1.567), Fig. 3.23. For this standard, the measured value is set to 100 gloss unit (GU).

We must remember that the light travelling into the wood surface is bent (refracted) in such a way that the ratio of the sine of the incident angle to the sine of the bent angle (see Fig. 3.23) is equal to the ratio of the refraction index of the wood to the refraction index of the air in which the light travelled before striking the interface.

Commercial measuring units are available with different measuring angles and opening area/measuring area. A bigger measuring area gives rather an average gloss value while a smaller measuring area (Φ 3 mm) is suitable to measure gloss patterns

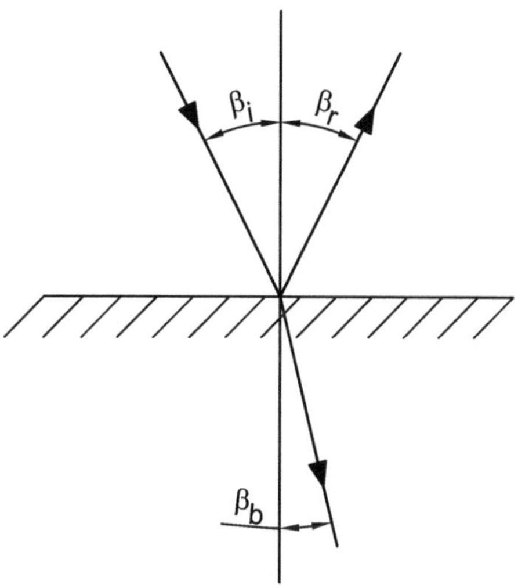

Fig. 3.23 Reflection and refraction of a ray on a surface

due to structural differences of wood surfaces. The latter may be used to characterise aesthetical values of high-demand wood surfaces.

In practice, there are different gloss grades to distinguish:

High gloss	70–100 GU
Gloss	40–70 GU
Silky gloss	25–40 GU
Silky mat	15–25 GU
Mat	less than 15 GU

Establishing gloss grades is always conditional, and therefore, the number and limiting values of gloss grades may vary considerably. New standards tend to classify only three grades such as gloss, semi-gloss and low-gloss. For a long time, the gloss grade "silky gloss" (Seidenglanz) has been the popular and preferred glossiness.

Early experiments have already shown that the apparent brightness is proportional to the directional emissivity or reflectivity (Schmidt and Eckert 1935; Eckert 1936). If the radiation intensity of a surface corresponds to a perfectly diffuse (hemispherical) radiation, then the directional emissivity or reflectivity is independent of the viewing angle and identical with the diffuse reflectivity. Actually, wood materials exhibit a certain degree of dependence of the reflectivity on the viewing angle depending on the local surface condition. A better radiating surface appears brighter when viewed under a larger angle and the opposite is valid for a poor radiating surface.

Due to the distinct structure (tracheids and vessels) of wood, the surface of a planed sample looks like a layer of cylindrical tubes. The optical scattering of the incident light strongly depends on the direction along or across the fibres. In the direction along the fibres, a combination of specular and diffuse reflection occurs while across the fibres a nearly diffuse reflection may be expected. Polishing the wood surface, the reflection ability will be enhanced in all directions, and therefore, also across the fibre a combined specular and diffuse reflection will take place.

It was also observed that the reflection of a surface is more directed for inclined incident light compared to perpendicularly incident light (Schmidt and Eckert 1935). This circumstance alone may cause difficulties in the correct interpretation of readings of different measuring angles. Therefore, it was decided to clear this question making comparative measurements on the same surface using different measuring angles (Sitkei 2014).

A wide-ranging investigation on surfaces with quite different glosses revealed that the GU_{20} and GU_{60} readings have a strong correlation in the entire range up to 100 GU. This correlation is represented in Fig. 3.24.

The good correlation may probably be explained by the fact that the incident angles 20° and 60° are large enough to be relieved of excessive influence of surface microstructure. The 85° incident angle is, however, much more sensitive to the surface roughness and grain angle.

It is interesting to note that the maximum GU_{60} readings on real wood surfaces amounted around 85 %. On a décor sheet, formerly extensively used on kitchen furniture, near 100 % could have been measured with a GU_{20} value of 90 %. The decor sheet has a very smooth surface, and therefore, the GU_{85} reading equals the GU_{60} reading, i.e. 100 %.

From the steep course of the correlation curve, we see that the specular component of the reflection is continuously increasing and at GU = 100 the reflection is more or less perfectly specular.

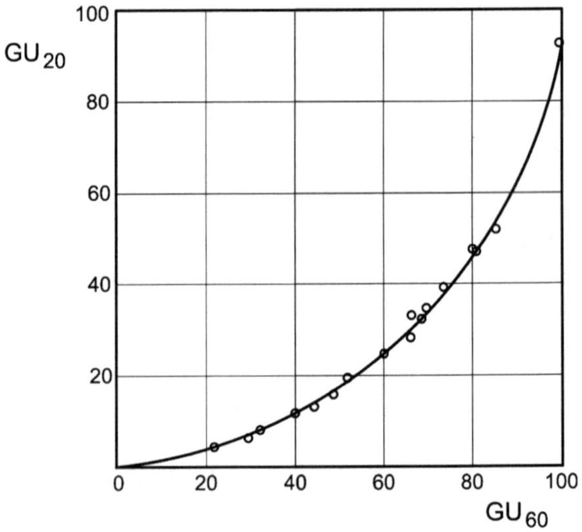

Fig. 3.24 Correlation between GU_{20} and GU_{60} readings for different wood surfaces

For the sake of a unified interpretation of the measured results, it is best to use the 60° measuring angle supplemented with the 85° angle (if needed) as will be shown later.

Measurements were undertaken on wood samples from all over the world and on different furniture including also quite old Biedermeier cabinet. Also some wooden artefacts, carvings and turneries, were also measured with their original surface finish (for example, linseed oil + shellac, beeswax + paraffine oil, and rice bran oil) Wood samples were machined, sanded, polished and surface treated with oil, wax or mix of beeswax + paraffine oil.

For gloss measurements, we used a gloss metre having all three standardised measuring angles (20°, 60° and 85°) with a measuring area of 8 × 14 mm. For local gloss measurements, a standard 60° unit was used with a small detecting area of Φ 3 mm. This was suitable to measure gloss differences especially on furniture surfaces. Readings of both units at the common 60° angle were comparable.

3.7.2 Measurement Results on Wood Samples

As outlined above, the reflection of wood surfaces is a very complicated phenomenon depending on many influencing factors. The main factors are the wood species, volume density, surface roughness, the direction of cut in relation to fibres, heart- or sapwood, earlywood or latewood, regular or irregular growth and surface treating (wax, oil, shellac, clear coat resin).

From the above-mentioned, it is clear that a very detailed experimental work was needed to answer the basic questions concerning gloss properties of wood surfaces. Preliminary experiments have already revealed that the gloss ratios measured with 85° and 60° incident light are distinct for a given surface and these ratios are always constant. Therefore, we illustrate the experimental results in a $GU_{85} - GU_{60}$ coordinate system. The measured points indicate, at the same time, the range of gloss on the given surface.

Figure 3.25 shows the correlation between GU_{85} and GU_{60} values for a planed Sugi cedar (*Cryptomeria japonica*) in fibre direction and perpendicular to the fibres. Coherent values measured on the same surface show a good correlation and the directional tangent always goes through the origin of the coordinate system. This means that the GU_{85}/GU_{60} ratios can be taken as constant for a given wood species and surface condition.

These ratios are, however, distinctly different in fibre direction and perpendicular to fibres. The occurrence of GU_{85}/GU_{60} ratios less than one requires some explanation. This happens always for more rough surfaces and with measurement perpendicular to the grains. Using high (flat) incident and viewing angle, the protruding parts of the surface (bumpiness) disturb the local reflection, and as a consequence, the GU_{85} readings show a lower gloss.

Using a subsequent sanding perpendicular to the grains, it considerably modifies the directional reflection of the surface and it results less differences in the gloss

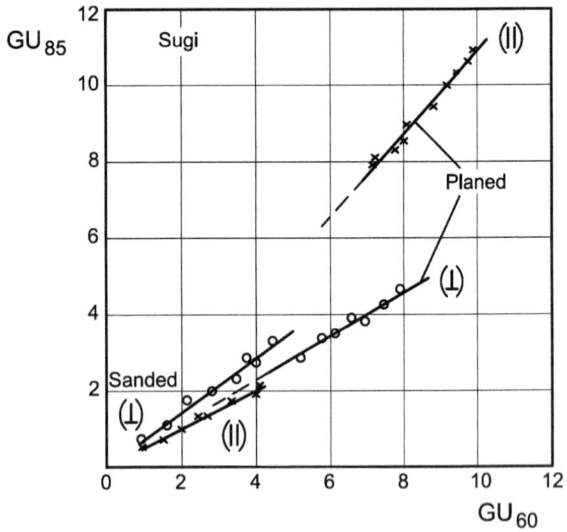

Fig. 3.25 Correlation between GU_{60} and GU_{85} readings, for planed and sanded surface. *Cryptomeria japonica*, $R_Z = 21 - 25$ μm, $\rho_v = 0.47$ g/cm^3

ratios compared to the planed surface. It is interesting to note that the sanding perpendicular to the grains apparently restructures the surface topography and higher gloss ratios are obtained along the scratches generated by the grits.

A more detailed picture will be seen in the following figures. Figure 3.26 shows measurement results on Sugi heartwood sample both on earlywood and latewood.

All characteristic parts have different local gloss, but the resultant gloss is, more or less, in agreement with Fig. 3.25, in despite of different samples. Similar measurement results are depicted in Fig. 3.27 for Sugi sapwood.

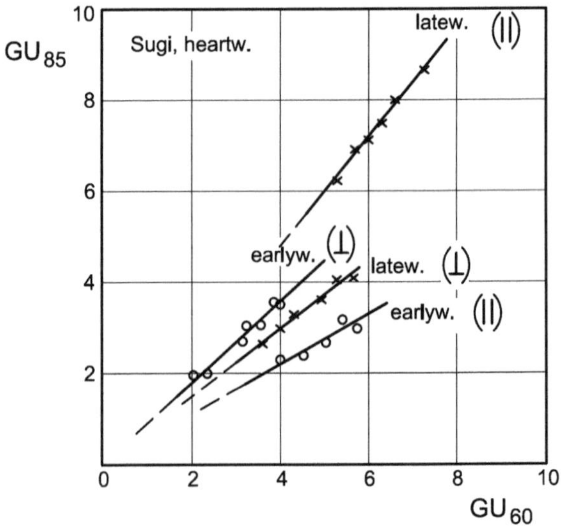

Fig. 3.26 Gloss ratios for Sugi heartwood (*Cryptomeria japonica*), $\rho_v = 0.43$ g/cm^3, $R_Z = 23 - 25$ μm

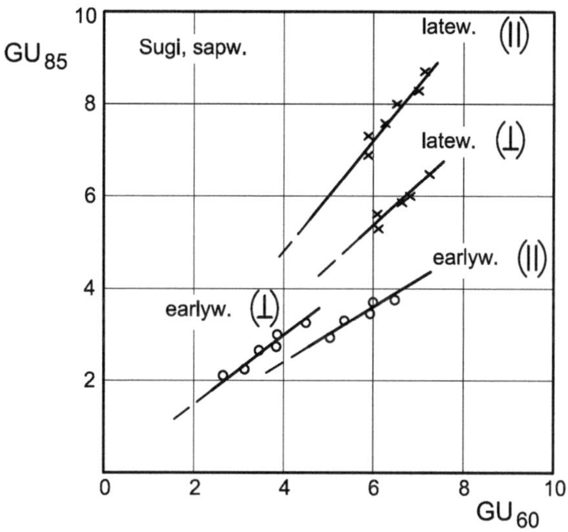

Fig. 3.27 Gloss ratios for Sugi sapwood (*Cryptomeria japonica*) $\rho_v = 0.43$ g/cm^3, $R_Z = 22 - 25$ μm

The earlywood is more bumpy, and as a consequence, readings with 85° angle have lower values. Furthermore, the GU_{60} readings perpendicular to the grains are always lower than those parallel to the grains. Latewood readings for 60° angle almost the same independently of grain directions, but their GU_{85}/GU_{60} ratios are different.

Figure 3.28 represents measurement results for Pomu sample (*Fokienia hodginsii*, $\rho_v = 0.54$ g/cm^3) which had somewhat higher volume density than the previous Sugi.

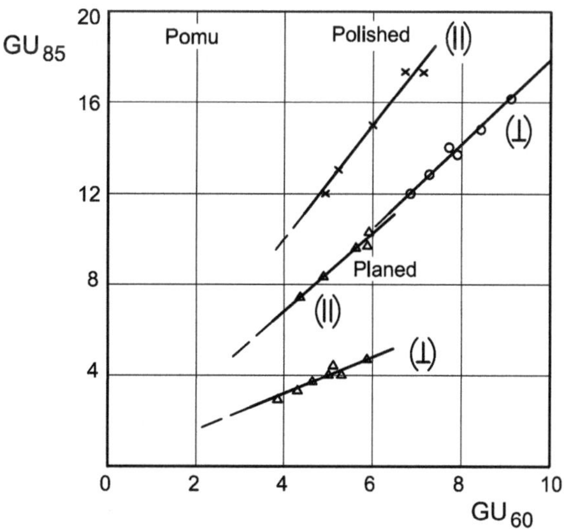

Fig. 3.28 Gloss ratios for Pomu (*Fokienia hodginsii*) with planed and polished surfaces

The planed surface shows somewhat higher ratios (1.7 and 0.81) but again they are distinct different from each other. The polished surface has higher gloss readings for 60° measuring angle, but also the gloss ratios are considerably increased. The difference in gloss ratios parallel and perpendicular to the grains remains however.

The gloss ratios of Redwood (*Sequoia sempervirens*, $\rho_v = 0.42$ g/cm^3) at different surface treatments are shown in Fig. 3.29. Reducing the surface roughness, the gloss values and gloss ratios are increasing and an additional surface treating with wax further enhances the gloss. The gloss ratios perpendicular to the grains have, however, low values.

The extreme fine grained Alerce (*Fitzroya cupressoides*) with P 400 sanding gives similar gloss values to those of Redwood with wax treating, Fig. 3.30.

Gloss ratios perpendicular to the grains, due to the smoother surface ($R_{pk} + R_k = 8.2$ μm), are higher than those we saw for Redwood. On the same figure, we see gloss ratios for a sugar box made of Tasmanian black-heart Sassafras (*Atherosperma moschatum*) polished with "hard timber finish". It has much higher GU_{60} values corresponding to a "silky gloss" grade. The wide range of glosses is also striking. The broad-leaved species show a similar picture but the higher density and hardness may affect the surface roughness and finish.

Figure 3.31 shows the gloss ratios of a planed Ash surface with a volume density of $\rho_v = 0.73$ g/cm^3 and $R_{pk} + R_k = 22.7$ μm. The gloss ratio in perpendicular to grains is comparatively low due to the higher surface roughness. A subsequent sanding perpendicular to the grains alters the gloss ratios, however, also here considerably.

When polished, the same species in polished condition (sanded with P 400, $R_{pk} + R_k = 8.2$ μm) show much higher gloss ratios, Fig. 3.32, but only moderately higher gloss readings with 60° measuring angle. The high gloss ratio perpendicular

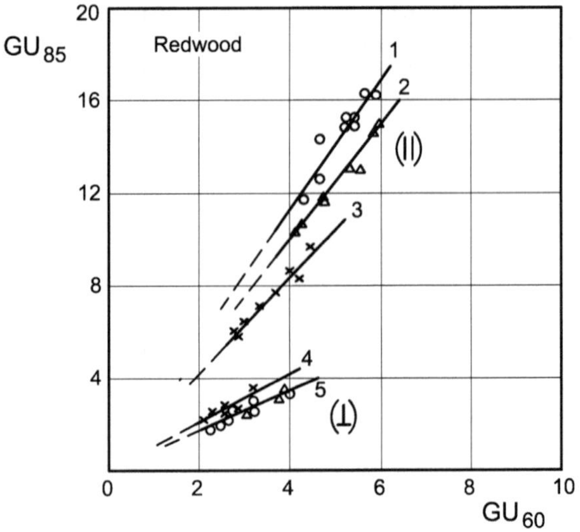

Fig. 3.29 Effect of different surface treatments on gloss ratios. Redwood (*Sequoia sempervirens*) milled (*3* and *4*), sanded P 400 (*2* and *5*), sanded and waxed (*1* and *5*), $\rho_v = 0.42$ g/cm^3

to the grains indicates a good surface roughness having no barrier to light reflection under large viewing angle.

A rougher sanding, even if the sample has a high density, gives always a lower ratio perpendicular to the grains. As an example, the measurement results on a Surian (*Toona sureni*, $\rho_v = 0.9$ g/cm^3, $R_{pk} + R_k = 12 - 13$ µm) are depicted in Fig. 3.33.

It was sanded with the grade P 240 giving $R_{pk} + R_k$ values between 12 and 13 µm. The gloss ratio perpendicular to the grains is only slightly over one. It is interesting to note that oil treating with rice bran oil has lowered the gloss values of GU_{60}, but the gloss ratio remained unchanged.

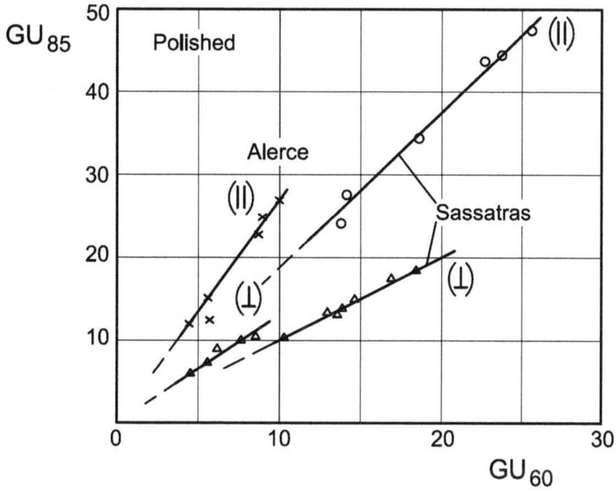

Fig. 3.30 Gloss ratios for polished Alerce (*Fitzroya cupressoides*) and Sassafras (*Atherosperma moschatum*)

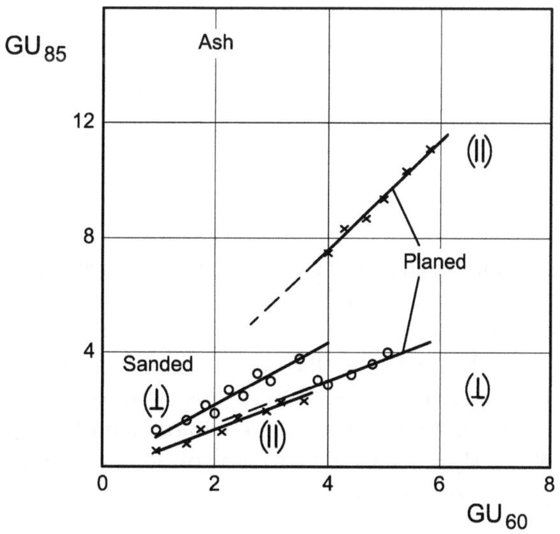

Fig. 3.31 Gloss ratios for planed and sanded Ash (*Fraxinus excelsior*), $\rho_v = 0.73$ g/cm^3

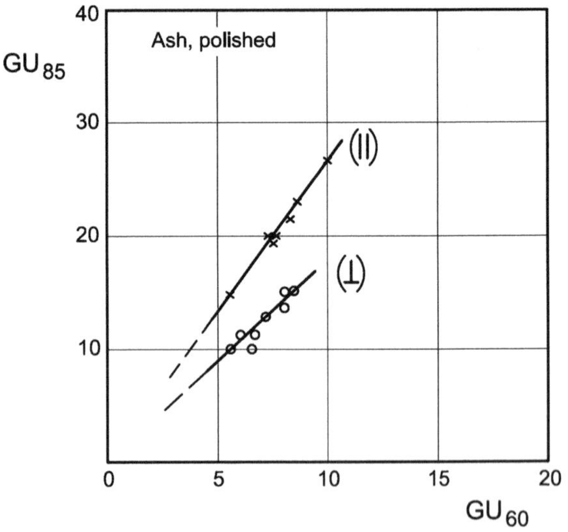

Fig. 3.32 Gloss properties of polished Ash (*Fraxinus excelsior*)

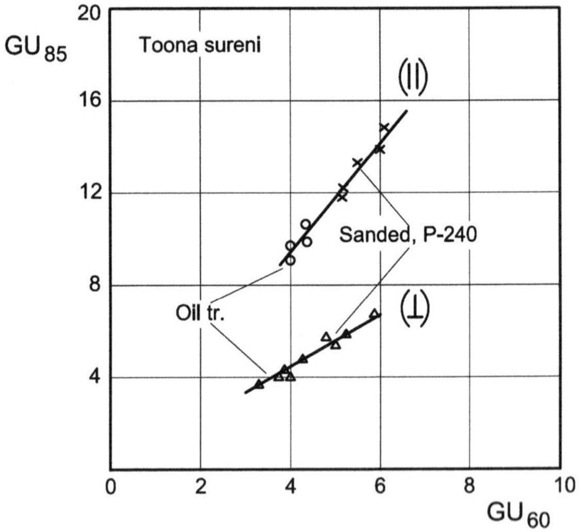

Fig. 3.33 GU_{85}/GU_{60} ratios for sanded samples with and without oil treatment (*Toona sureni*)

Quite similar, gloss ratios were obtained on a *Parashorea chinensis* sample having roughness values of $R_Z = 42 - 45$ μm, $R_{pk} + R_k = 10.5 - 13$ μm and $\rho_v = 0.83$ g/cm^3.

Further increasing the quality of finish, or more precisely, lowering the roughness value of $(R_{pk} + R_k)$ the gloss ratio is continuously increasing. Figure 3.34 shows measurement results on Karri sample (*E. diversicolor*), $\rho_v = 0.92$ g/cm^3, $R_{pk} + R_k = 7$ μm) giving ratios of 3.5 and 2.0 parallel and perpendicular to the grains, respectively.

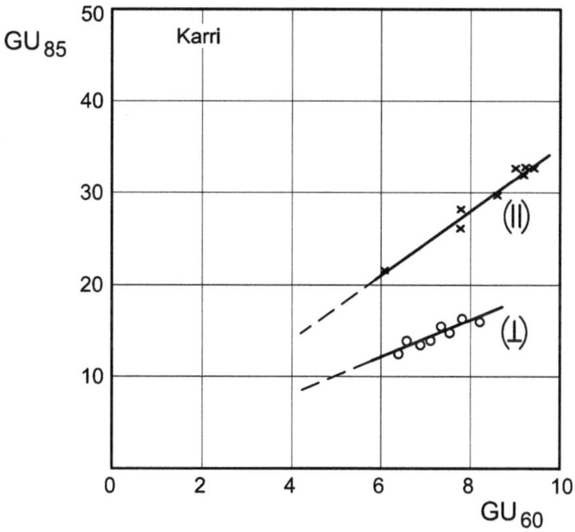

Fig. 3.34 Gloss ratios for Karri wood (*Eucalyptus diversicolor*)

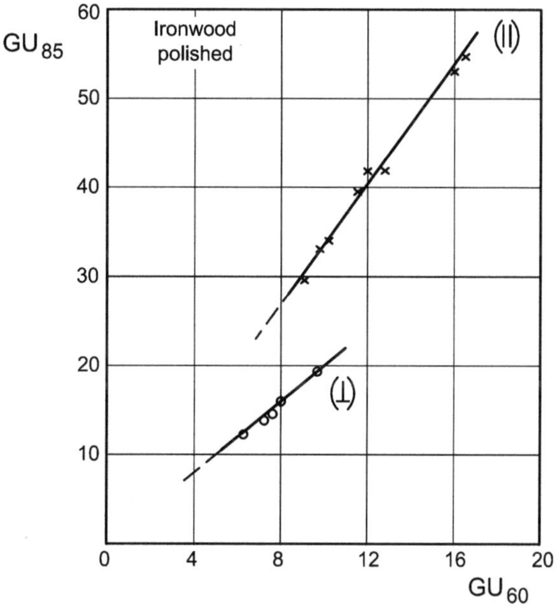

Fig. 3.35 Correlation for GU_{60} and GU_{85} readings. *Olea capensis*, $\rho_v = 1.02$ g/cm^3, $R_{pk} + R_k = 6.1$ µm

The South African Ironwood (*Olea capensis*, $\rho_v = 1.02$ g/cm^3, $R_{pk} + R_k = 6.1$ µm) gave quite similar results independently of the difference in tree families, Fig. 3.35.

A wax treatment (beeswax + paraffine oil) widened the gloss values, especially parallel to the grains, but the gloss ratios remained unchanged.

Ultimately let we see the measurement results on a very dense Southeast Asian Rosewood or Palisander (*Dalbergia cochinchinensis*, ρ_v = 1.12 g/cm^3, R_Z = 11 – 14 μm, $R_{pk} + R_k$ = 3.6 − 3.9 μm) which illustrated in Fig. 3.36.

The gloss ratio is extreme high, also perpendicular to the grain. These findings suggest that a given correlation may exist between the gloss ratios and the $R_{pk} + R_k$ roughness parameter. Using our measurement results, Fig. 3.37 shows the expected

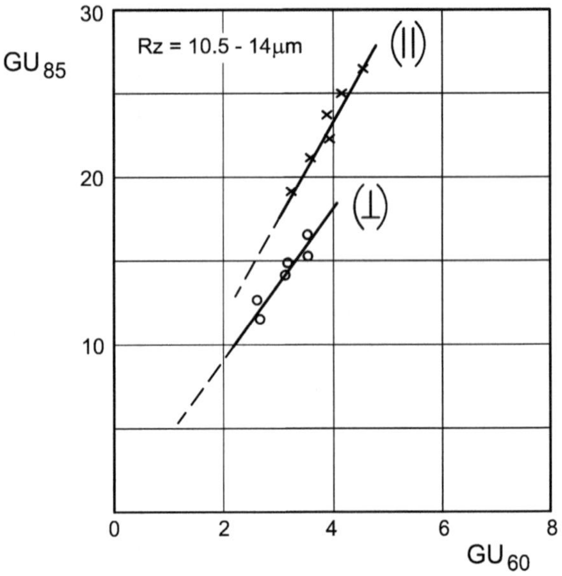

Fig. 3.36 Gloss properties of an extreme dense Rosewood (*Dalbergia cochinchinensis*)

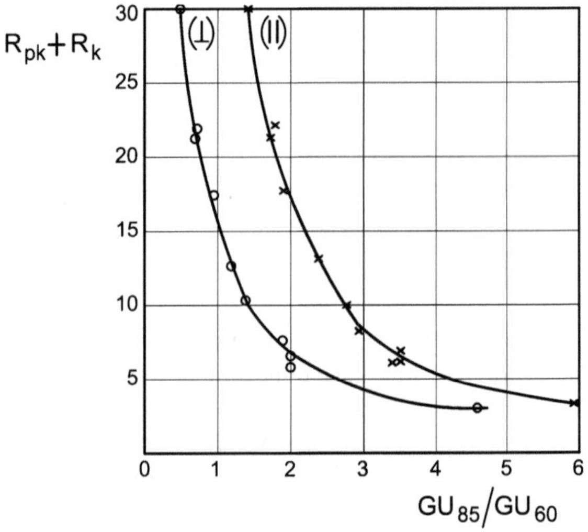

Fig. 3.37 Correlation functions between roughness parameter (R_{pk} + R_k) and gloss ratios parallel and perpendicular to the grains

correlation functions between the roughness parameter (R_{pk} + R_k) and the gloss ratios parallel and perpendicular to the grains.

This correlation is valid first of all in cases, when the dominant effect influencing the reflection is originated from the surface structure of wood. This constrain is fulfilled by natural wood surfaces without any treatment or treated with wax, oil or shellac, even in thin layer. Different resin clear coatings depending on the coating thickness may alter the reflection properties considerably. Furthermore, irregular growth structures may diminish or even cease the difference between the gloss ratios parallel and perpendicular to the grains. At the same time, many veneered surfaces treated with thin stain or clear resin, more or less, correspond to the above correlation.

3.7.3 Measurement Results on Furniture

In the following, experimental results measured on veneered furniture surfaces will be presented. Figure 3.38 illustrates the gloss ratios of an oak-veneered cabinet in natural colour (thin clear coating) manufactured in the 1970s. The gloss parallel to the grains varies in wide range corresponding to the silky mat grade. Both gloss ratios are in agreement with the generally valid correlation functions given in Fig. 3.37.

Figure 3.39 shows the gloss ratios of a big beech-veneered sound box slightly stained to a middle brown colour and manufactured in the 1960s.

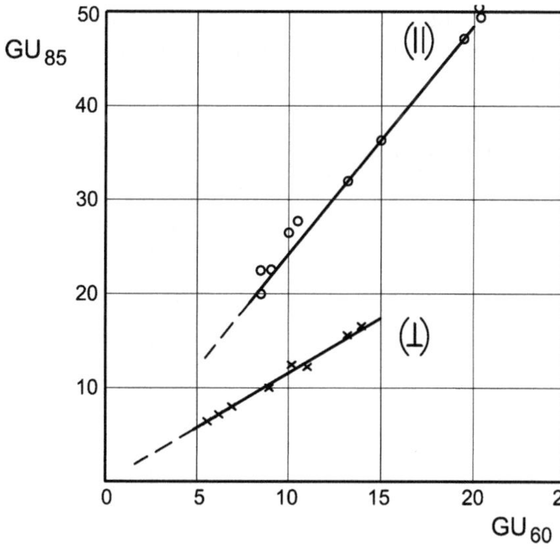

Fig. 3.38 Gloss ratios of an oak-veneered cabinet in natural colour

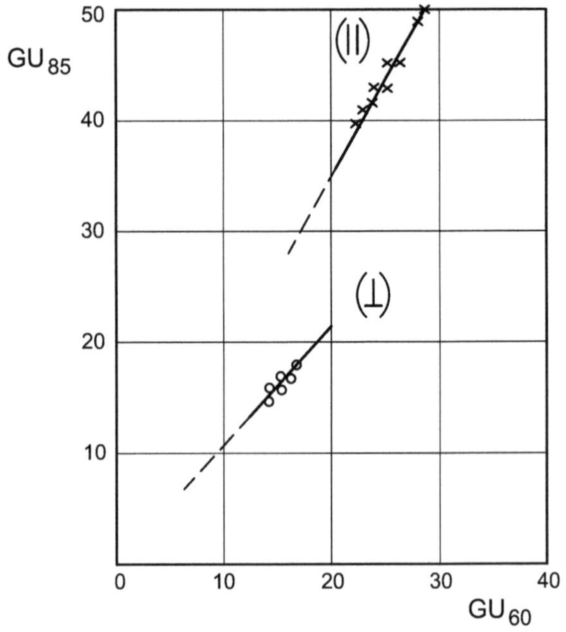

Fig. 3.39 Gloss properties of a beech-veneered sound box stained to middle brown colour

The gloss changes in a narrower range corresponding to the silky mat grade. The gloss ratios are almost in agreement with the general correlation functions (Fig. 3.37).

Figure 3.40 illustrates the gloss ratios for a beech-veneered book shelf which was seemingly not sanded to smooth but stained to a deep brown colour. The gloss parallel to the grains corresponds to the upper boundary of the silky mat grade. Due to the rough surface, the gloss ratio perpendicular to the grains is less than one.

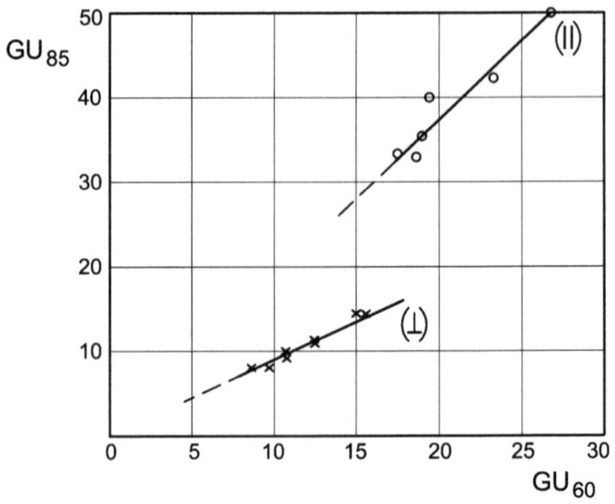

Fig. 3.40 Gloss ratios of a beech-veneered book shelf

The gloss ratios also here more or less correspond to the correlation curves shown in Fig. 3.37.

The gloss ratios of a walnut-veneered table stained to a middle brown colour are depicted in Fig. 3.41.

The gloss parallel to the grains varies in a broad range corresponding to the silky mat grade. The gloss perpendicular to the grains is changing in a narrower range and its ratio is somewhat higher than that would correspond to the correlation function.

A bed frame of solid pine in natural colour was also measured, manufactured in the late 1970s, and the gloss ratios are illustrated in Fig. 3.42.

The measured surfaces more or less correspond to the tangential cut and here the gloss ratios are in agreement with the correlation functions given in Fig. 3.37. Local measurements near the plank edge have shown, however, a higher scattering. A closer look at the surface revealed that the planks were cut from logs of small diameter, and therefore, near the edge of the planks the direction of cut is turning radial. This example should draw our attention to the visual inspection of the given surface in question. All kinds of texture variations may cause distortions in the measured gloss values.

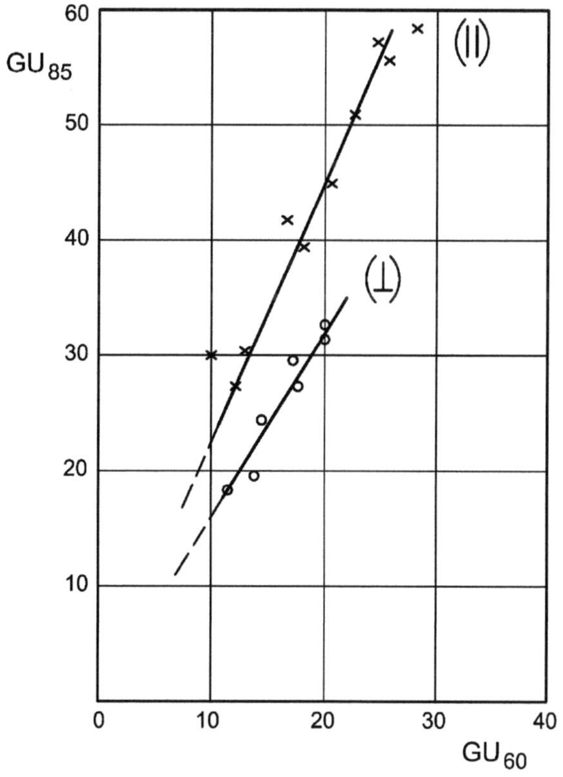

Fig. 3.41 Gloss measurements on a walnut-veneered table slightly stained to middle brown colour

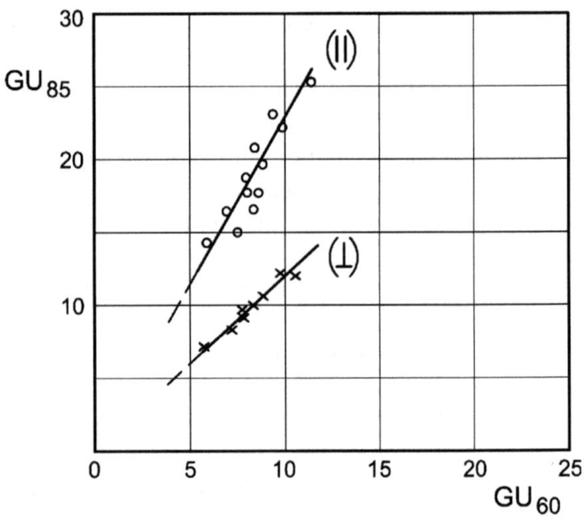

Fig. 3.42 Gloss measurement on a solid pine furniture

Measurements were also performed on an old Biedermeier cabinet, manufactured well over 100 years ago, with original surface finish. Walnut wood has been used and finished with shellac (Politur) in several subsequent layers. This comparatively thick layer hides the fine details of surface texture and only the strikingly marked figures and the differences in colour hues will be effective. The measurement results are depicted in Fig. 3.43.

The gloss values are the highest we have measured on any kind of furniture and it corresponds to the high gloss grade. The GU_{60} and GU_{85} readings differ only slightly from each other independently of grain directions. The gloss readings

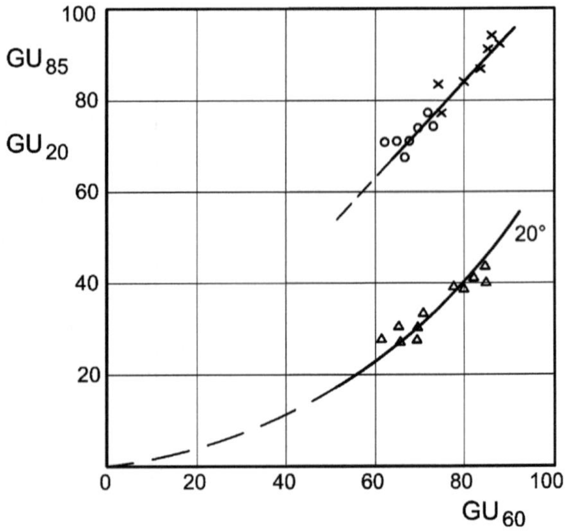

Fig. 3.43 Gloss readings on an old cabinet using all three measuring angles. (*x*) Parallel to the grains, (*o*) perpendicular to the grains, (Δ) 20°, mixed grain angles

perpendicular to the grains are, however, somewhat lower. The readings at 20° viewing angle are also given and they correspond to the course given in Fig. 3.24.

In order to save the high-quality timber with good mechanical and aesthetical properties, the three-layer strip flooring is widespread. The top layer may be cut in different directions in relation to the bole main axis, and therefore diving fibres may or may not occur. It is very common that the top layer pieces are cut mixed between the pure tangential and radial directions. These surface texture differences may be reflected in the gloss measurements made on several top layer pieces.

Figure 3.44 shows gloss ratios of strip flooring having Ash top layer and clear coated with acrylic resin.

The measured surfaces had no diving fibres. The coated surface is relatively smooth $R_Z = 24 - 26$ μm, $R_{pk} + R_k = 8.5 - 9.5$ μm and the dips are originated from the cut of vessels. The gloss ratios are corresponding to the generally observed correlations given in Fig. 3.37. Gloss readings at 60° measuring angle are practically the same in all directions. That means that an observer using only the 60° measuring angle cannot recognise any differences in the reflection properties of this surface.

The measurement results of oak strip flooring clear coated with a high gloss resin are given in Fig. 3.45. The surface has diving fibres and it is highly reflected in the gloss ratios. The ratios are not far from each other, but the 60° readings parallel to the grains are considerably higher than those measured perpendicular to the grains. The readings for 20° measuring angle are also given and they follow the same course obtained in Figs. 3.24 and 3.43; the gloss values correspond to the "gloss" grade.

Further measurements on strip flooring (mainly tangential cut) are summarised in Table 3.2.

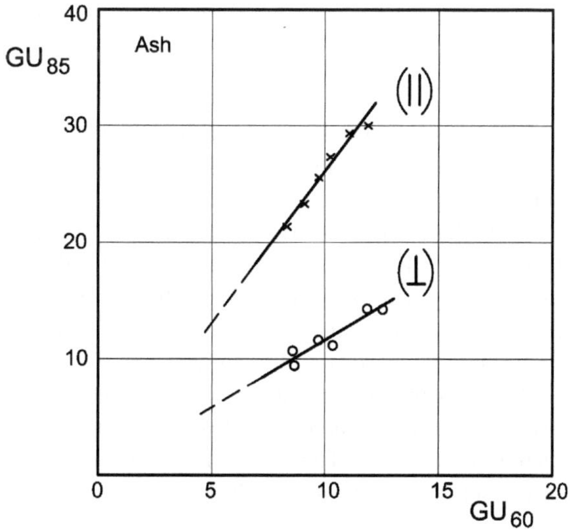

Fig. 3.44 Gloss measurements on a strip flooring with Ash top layer

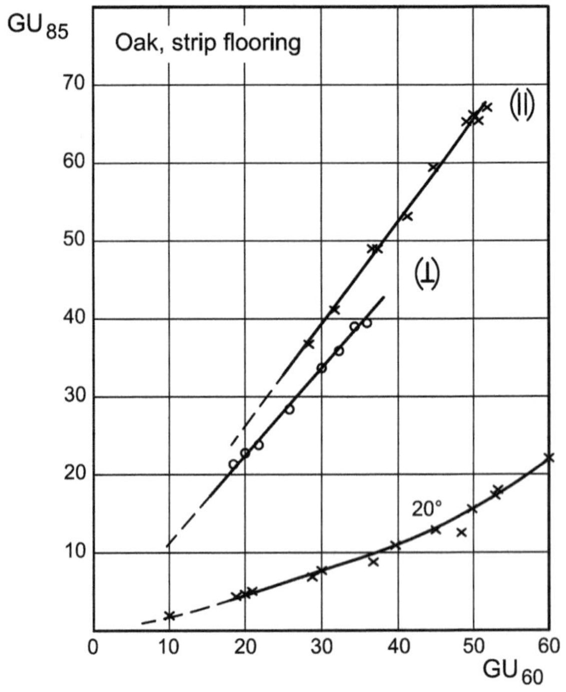

Fig. 3.45 Gloss ratios for oak strip flooring clear coated with high gloss resin

Table 3.2 Gloss properties of strip flooring materials

Wood species	Maximum gloss, 60°		Gloss ratios, 85°/60°	
	(‖)	(⊥)	(‖)	(⊥)
Guapinol *Hymenaea courbaril*	20	13	1.5	0.7
Merbau *Intsia palembanica*	20	13	1.45	0.55
Doussie *Afzelia rhomboidea*	20	13	1.65	0.62

The measurement results show that the wood species have little influence on the gloss properties. The resin used for clear coating determines the gloss grade of the surface considerably.

A similar three-layer construction can also be used for decorative wall panelling. In this case, the surface is not subjected to abrasive loading so it has a surface treating only with oil. Due to the high aesthetical requirement, the top layer sheets are carefully cut giving the maximum achievable figure pattern beauty, which may be further enhanced by oil treatment.

The measurement results are depicted in Fig. 3.46. In contrast with the results given in the above table, the gloss ratios are highly different parallel and perpendicular to the grains. Therefore, the grains are clearly accentuated and gentle figures of the surface appear in its full beauty.

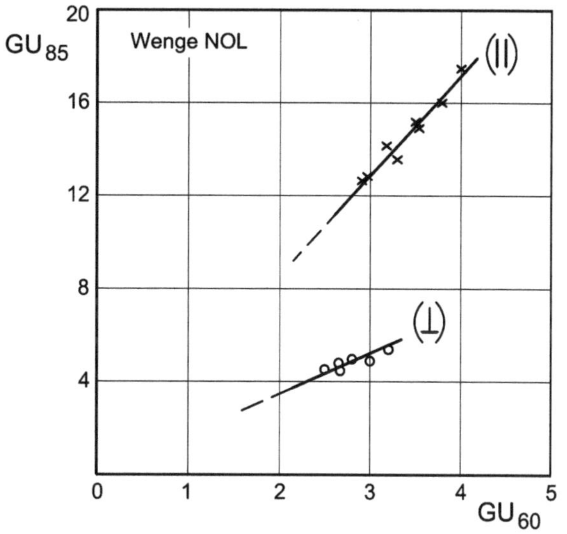

Fig. 3.46 Gloss ratios of a wood surface for decorative panelling, tangential cut, with oil treatment (*Milletia laurentii*). *Colour* blackish dark brown

Similarly to Figs. 3.26 and 3.27, the latewood having a pitch-black colour has considerably higher gloss values in both directions compared to those of the earlywood. Figure 3.46 represents rather the average values measured on a larger surface.

The characteristic feature of many timber species is their irregular growth structure which may bestow an exceptional beauty to wood products. Such irregular growth structures are the burls, bird's eye, blotched, mottled and streaked structures, curly wood, fiddle-back, etc. As a rule, an irregular growth pattern is also accompanied by distinct colour changes and they further enhance the beauty and value of the products further.

The visible figure pattern fundamentally depends on the dominant structural elements giving the distinct appearance to the surface and the direction of cut. In general, the tangential cut reveals the distinct grain pattern in texture and colour. In other cases, especially in many tropical species, the growth rings are not well-defined and the attractive structure and colour are given by very prominent dark-coloured rays (ray figure). As an example to the latter, the New Zealand Rewarewa (*Knightia excelsa*), depending on the direction of cut, may show its quite different face: a pure radial cut shows large (20–30 mm) dark brown blotches evenly distributed on the middle brown base. Turning the plane of cut towards tangential, the blotches are continuously diminishing and the surface pattern will be more mottled. Finally, at tangential cut produces a fine streaked surface is obtained. The Beefwood (*Grevillea striata*) shows its very characteristic figure pattern only near a radial cut when the plane of cut is inclined about 10°–20° to the radial plane.

Figure 3.47 shows gloss ratios parallel and perpendicular to the grains for two wood species with strongly irregular growth. The grains have a more or less curly character and a random direction in relation to the main axis.

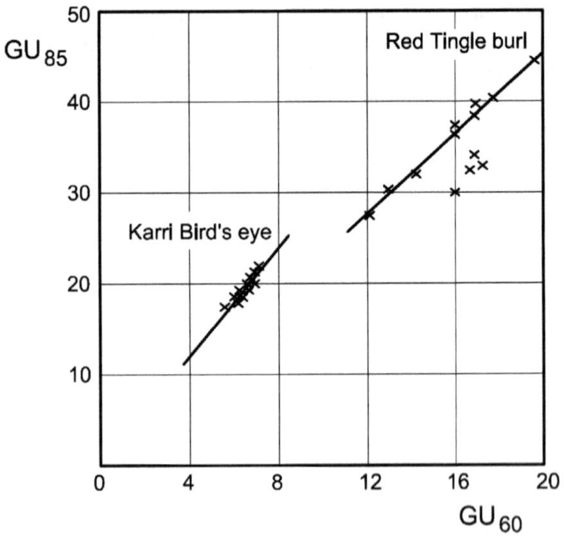

Fig. 3.47 Reflection properties of wood species with irregular growth. Red Tingle = *Eucalyptus jackmanii*; Karri = *Eucalyptus diversicolour*

The gloss ratios have considerable scattering, especially for the Red Tingle burl, but they have no definite difference independent of directions.

A seemingly irregular growth is not accompanied by irregular fibre structure in its entire volume. In such cases, the surface behaviour follows the normal behaviour of wood species with regular growth. Figure 3.48 represents the measured gloss for a Karri bird's eye (*E. diversicolor*) in the main direction. The surface is polished and treated with a mix of beeswax and paraffine oil. The average roughness was $R_Z = 16$ µm and $R_{pk} + R_k = 6.1 - 6.7$ µm. The scattering of measurement points is also here slightly higher. The difference between the two ratios, however, corresponds to the generally valid correlation curves given in Fig. 3.37.

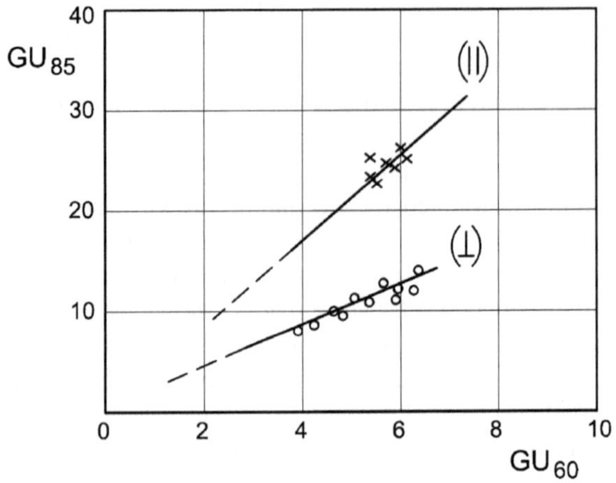

Fig. 3.48 Gloss ratios for Karri bird's eye with regular grain structure

Some wood species are inclined to form irregular growth also within their bole, especially just over the roots accompanied with colour changes. For example, the walnut bole may have the diversity of structure and colour which was highly utilised formerly in making decorative cabinets and other furniture (the age of Biedermeier).

Figure 3.49 shows the gloss ratio for a walnut veneer with irregular growth used on a bed-clothes cabinet. Due to the higher average glossiness, the gloss ratio is not too high (1.6) similarly to those given in Fig. 3.45. At the same time, no difference occurs when measuring from different directions.

Finally, we examine the reflection properties of a speciality from the tree-kingdom, the timber of Holly-leaved Banksia (*Banksia ilicifolia*) having beautiful dark red wood made into a kitchen-board, Fig. 3.50.

The visible structure is similar to that as if the fibres would be woven into a loose web in the form of thin whitish lines. The gloss ratios are only slightly different from each other. The surface is smooth planed and oil treated.

From a certain point of view, the man-made medium dense fibreboard (MDF) has a quite different texture compared to natural wood materials. It may be treated here among the timbers with irregular growth structure. The MDF is lacking well-ordered fibre structure, and therefore, one would expect no dependence of gloss ratios on the direction of measurement. However, after pressing MDF boards will be equalised to a given thickness using rough sanding. The marks of the sanding particles on the surface act somewhat similarly to the grains, and as a consequence, an MDF surface without finishing shows a similar picture to that of natural wood surface. Figure 3.51 gives measurement results on a raw MDF board (ρ_v = 0.84 g/cm^3).

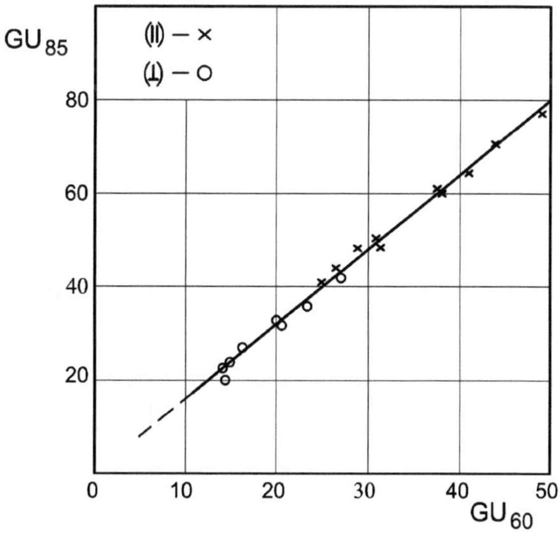

Fig. 3.49 Gloss properties of a walnut veneer with irregular growth

The gloss ratios differ from each other quite similarly to wood surfaces. After removing the marks of the rough sanding, the gloss ratios become much the same in all directions, although the scattering of measurement points is higher.

Based on the vast amount of experimental results, some general tendency in the variation of gloss ratios can be formulated. The following rule of thumb can be

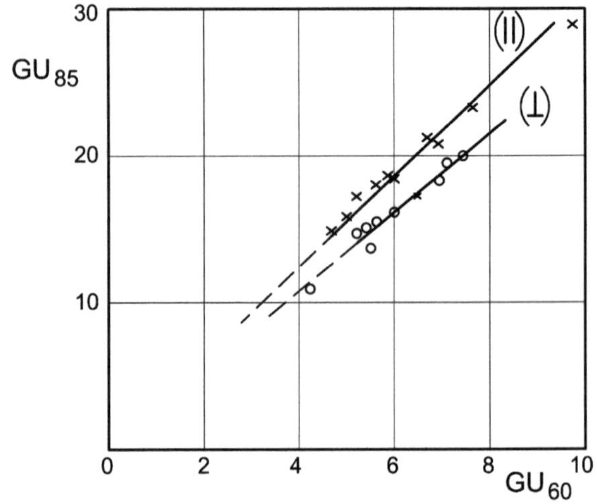

Fig. 3.50 Reflection properties of a holly-leaved Banksia with irregular growth

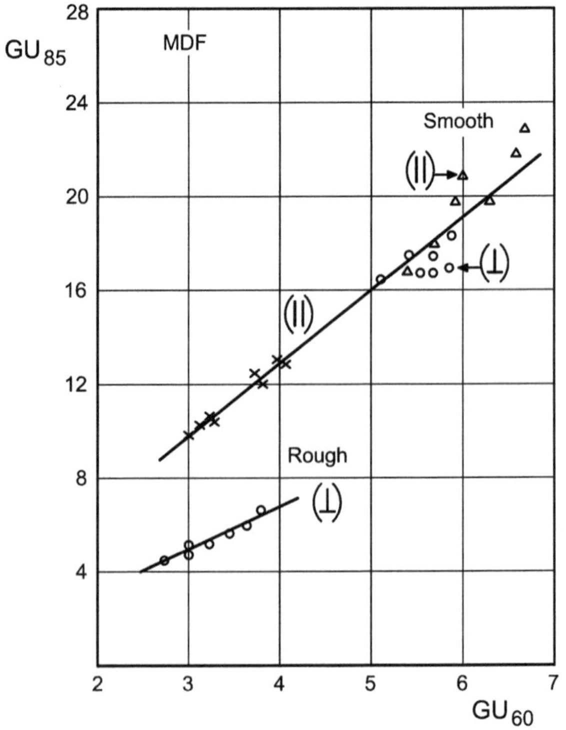

Fig. 3.51 Gloss properties of a raw MDF board and after finishing

established for gloss measurements parallel to the grain. The maximum gloss ratios vary between 5 and 6 providing a very smooth surface and gloss readings $GU_{60} < 20$. As the gloss readings increase, the expected gloss ratios will decrease quickly and they tend towards one as the gloss readings approach 100. At low gloss readings, the surface roughness plays a decisive role and here a very broad range of gloss ratios may be expected.

Figure 3.52 shows the field of experimental results. The limit value of one follows from the fact that high gloss values around 100 can be obtained by high gloss surface coatings giving a smooth surface without any orientation properties.

Based on theoretical considerations and experimental results performed with wide-ranging wood species and types of furniture, the following main conclusions may be drawn:

- due to the specific structure of wood surfaces, its light reflection properties depend on many factors and on a given surface area different reflection values can be measured,
- the gloss properties of a wooden surface depend on both the incident and the viewing angle; therefore, a unified approach and measurement method are essential,
- for gloss measurement on wood surface, in general, the 60° incident and viewing angle may be recommended, although the use of one measuring angle can give only limited information,
- the determination of GU_{85}/GU_{60} ratios is also recommended to gather more information. This means a measurement with two incident and viewing angles,
- the GU_{85}/GU_{60} ratios are generally constant, and the scattering of gloss ratios depends on the homogeneity of surface structure,

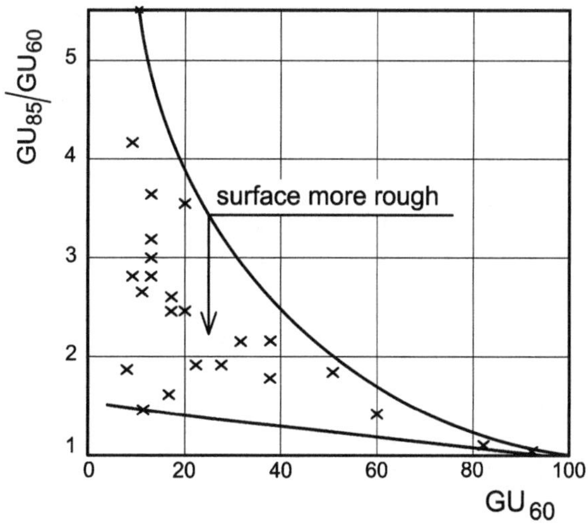

Fig. 3.52 The field of experimental results

- for normal wood surfaces, there is a distinct difference in the gloss ratios for directions parallel and perpendicular to the grains. Using one measuring angle, a lot of information is not attainable,
- the gloss ratio for directions parallel and perpendicular to the grains can be modified by breaking the continuity of reflecting mirrors using a proper surface finish,
- local gloss properties on a given sample, i.e. earlywood and latewood, may differ considerably,
- the GU_{20} readings are generally very low and they are not suitable to distinguish among gloss properties of wood surfaces. Nevertheless, GU_{20} readings are in correlation with GU_{60} readings,
- the gloss properties of uncoated or clear coated surface with wax, oil and thin resin layer correlate well with the surface roughness parameter (R_{pk} + R_k),
- shellac and clear coating resin may hide the fine structure of wood depending on their glossiness,
- irregular growth structures show considerable deviations compared to normal wood structures,
- the range and field of expected gloss ratios are highly depending on the average gloss value. As the average gloss approaches 100, the gloss ratio tends to one.

3.8 Practical Applications

The purpose of this section is to demonstrate the existing experimental materials to fulfil the practical demands of the wood industry and to give general explanations for the different discolouration process. Because a practising engineer is mainly interested, first of all, in the knowledge and identification of lightness and colour hue, one of the most important tasks is to connect the measured colour characteristics and the colour hue of a given wood sample. For these purpose, several new methods and representations will be demonstrated.

3.8.1 *Effect of Machining on Colour and Gloss*

The machining of wood always has special purposes such as cutting to a given rough size, giving exact shape and measure, to ensure the prescribed surface quality expressed in both physical and aesthetical parameters. Due to the anisotropic structure of wood, the method of cutting a bole is important to ensure the required properties in surface roughness, grain pattern, ray figures and colour combinations producing the maximum possible aesthetic values.

Sadly, in our times, the portion of small logs is continuously increasing, and therefore, a careful selection of raw materials and their purposeful cutting up is very important task.

Due to the mechanical anisotropy and brittle behaviour of wood, a pure cutting is never possible. The round edge of the tool always causes permanent deformations and brittle fractures in the surface layer of a given thickness depending on the edge radius and material properties. The brittle fracture of cell walls generally bestows a grey veil settling on the given colour hue which can be removed or diminished by applying a surface treatment with wax, oil or other clear coats. As outlined in Sect. 3.2, the presence of grey veil can be observed in the reflection spectrum whenever an increased reflection occurs in the 400–500 nm wavelength range. A proper treatment diminishes the reflection, first all, in this very range.

A surface finish, especially combined with oil or wax treatment, fundamentally diminishes the grey veil on the surface and the colour hue will be more vivid. At the same time, the lightness is always slightly decreasing. As an example, Fig. 3.53 shows reflection spectra for Kauri wood in natural and polished conditions (Sitkei 2013). Samples with a polished surface have practically no reflection in the 400–500 nm range.

Using different machining process, the microrelief of the created surface will also be different and it can show slightly variable colour hues. As an example, Fig. 3.54 shows reflection spectra of Scots pine sample after smooth planning and sanding with P 120. The sanding due to its distinct cutting mechanism with negative rake angle creates a surface enhancing the light scattering considerably and slightly modifies the colour hue towards grey.

The gloss of wood surfaces is more dependent on the surface roughness than their colour hue (see Fig. 3.37). At the same time, surface microstructure of wood and the attainable surface roughness is also a function of volume density of wood. Therefore, it is interesting to examine a possible correlation between gloss ratio and volume density. A selection of experimental results is depicted in Fig. 3.55 showing

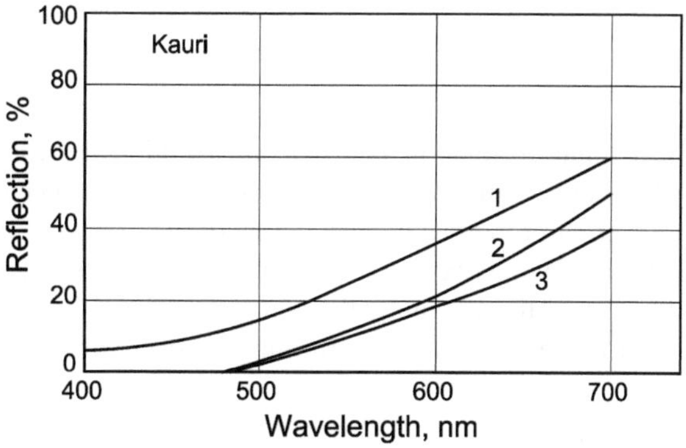

Fig. 3.53 Reflection for Kauri wood (Agathis australis). (*1*) Machined, (*2*) polished, (*3*) swamp Kauri polished

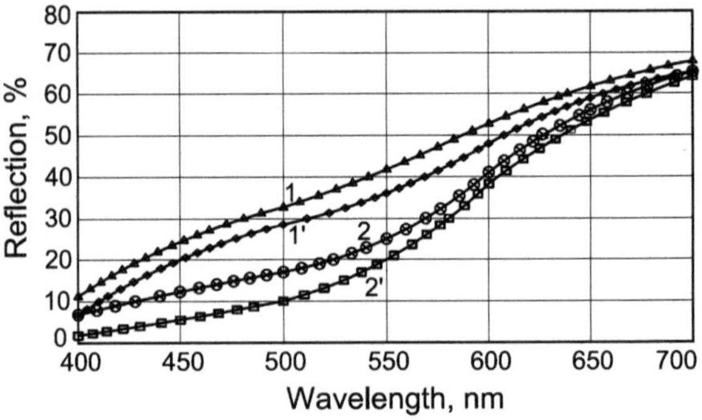

Fig. 3.54 Reflection of a Scots pine sample after planing and sanding with P 120. (*1,1'*) Earlywood planed and sanded, (*2,2'*) latewood planed and sanded

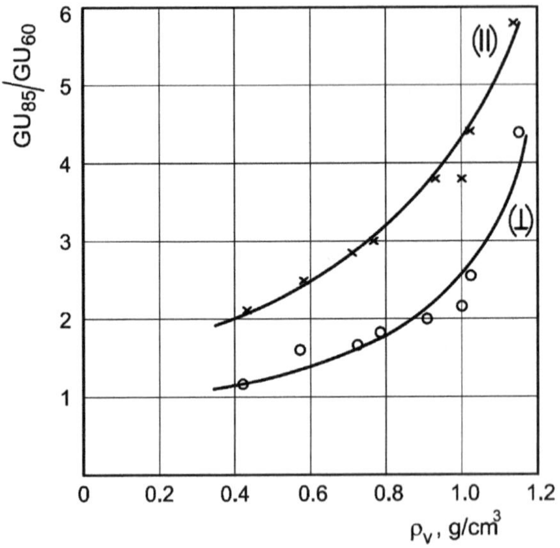

Fig. 3.55 Gloss ratios for polished samples as a function of volume density

a definite correlation between the mentioned variables. It is clear that also the characteristic microstructure of particular wood species has an influence on the light scattering properties independently of their density. For example, whether a conifer or a hardwood species is concerned with the same volume density. Nevertheless, the gloss ratios are uniquely increased, at least three times, in the density range from 0.4 to 1.15 g/cm^3 and confirming the overwhelming priority of volume density.

3.8.2 Enhance of Colour

The brittle fracture of cell walls in machined wood surfaces almost always produces a mat-coloured appearance with a greyish veil. Different colour enhancing methods can be used to produce a more distinct and vivid colour hue.

The simplest and most often used procedure is a surface treatment with natural (vegetable) resins, gums, oils, wax and lacquers or their mixture with shellac. It should be noted that shellac is an insect secretion that is formed on and collected from plants. Shellac is produced almost entirely in India (written memories since 1590), but a smaller quantity comes from Myanmar and Thailand. Shellac may be combined with Kauri resin and coloured by the incorporation of pigments. It is a spirit varnish and yielding a tough film with a smooth finish resulting in a high polish with GU_{60} = 80–90 %.

The addition of Kauri resin to shellac gave better water resistance and greater elasticity to the coating. The latter proved to be very important to avoid fine cracks and fissures by the flight of time. After 150 years, a Biedermeier cabinet keeps its full beauty in colour and gloss. Therefore, it is no wonder that fine furniture makers over Europe have widely been used this surface finish. Today, several leading wood turners in New Zealand use the mix of linseed oil and shellac claiming that shellac will ensure the gloss for many years.

The optical mechanism for enhancing colour can be examined and explained using the reflection spectrum, especially in the wavelength range of 400–550 nm. Experimental results with Karri wood (*E. diversicolor*) are depicted in Fig. 3.56 in its natural condition, or treated with beeswax and teakoil. After treating, in both cases, the reflection in the 400–500 nm range decreased considerably. The change in colour hue (λ_0 or h^*) is insignificant but the lightness L decreased by 30 %, while the saturation of colour C increased about 40 %. This fundamental increase in colour saturation will be perceived by observer as a *colour enhancement*. Physically, the oil or wax treatment lowers or removes the reflection in the lower wavelength range. By removing the grey veil, the colour appears more vivid.

Polishing the wood surface, the broken cell structure will be transformed into a more or less smooth surface which also reflects much less in the lower wavelength range. We have demonstrated results in Fig. 3.53 for Kauri wood. The machined surface has considerable reflection in the 400–500 nm range. After polishing, however, we see almost no reflection in this range, but the saturation of colour increased by 60 % while the lightness decreased by 30 %. The colour hue changes very little.

Figure 3.57 shows reflection spectra of a dark, almost black wood (*Milletia laurentii*) as a raw material and after finishing to a decorative panelling. The raw material has a scarcely visible brownish tone filtering through the blackish base colour. The measured colour hue is $h^* = b^*/a^* = 1.77$ or $\lambda_0 = 581$ nm.

After finishing and treating with a hardening oil, the surface lost its brownish tone and became almost pitch-black with $h^* = 0.43$ and $\lambda_0 = 610$ nm. Looking at the reflection curve 2, in Fig. 3.57, it has low reflection up to the 600 nm wavelength. Therefore, the standard lightness L^* determined only by the Y (green) tristimulus

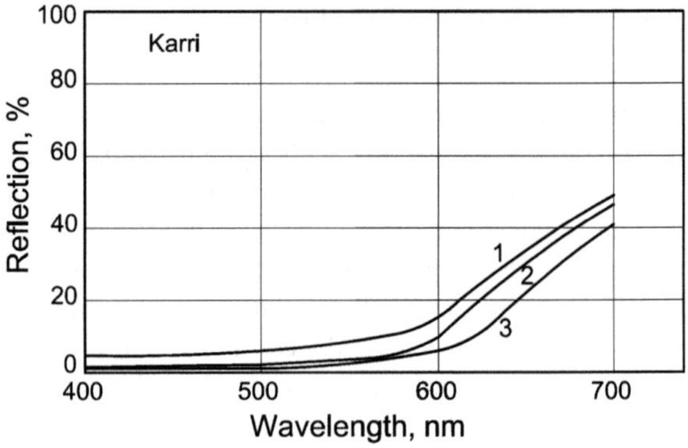

Fig. 3.56 Influence of surface treatment on the reflection of Karri wood, *Eucalyptus diversicolor*. (*1*) Natural surface, (*2*) wax, (*3*) teakoil

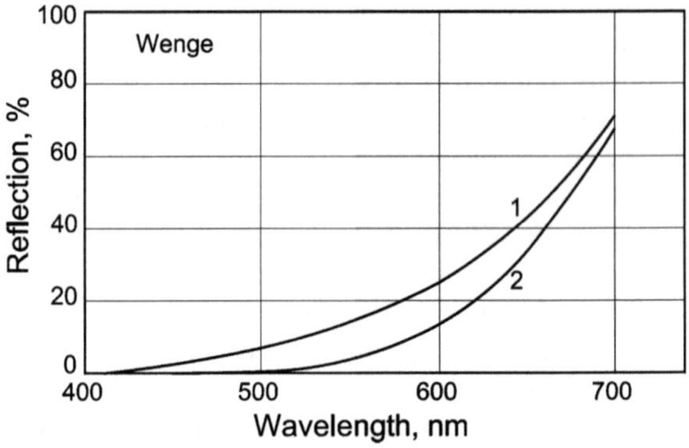

Fig. 3.57 Reflection of raw material (*1*) and after finishing (*2*). *Milletia laurentii*

value is disproportionately low. The standard lightness ratio is $L_1^*/L_2^* = 13.4/4.3 = 3.1$ while using the physical lightness $R_{m1}/R_{m2} = 2.086/1.453 = 1.436$. The latter ratio much more corresponds to the perception of the naked eye.

It should be noted that between the lightness reduction and increase in colour saturation, no correlation was observed. At the same time, some correlation was found between lightness reduction and colour hue of different wood species due to rice bran oil treatment. This correlation is given in Fig. 3.58 for many wood species. The correlation is weaker for reddish coloured species.

This finding may be related to the fact that even in this range, h^* determines the colour hue not very exactly. Using λ_0 as a hue characteristic, there is better hue determination.

The physical lightness R_m and its relative value shows a good correlation with the standardised lightness L^* which is depicted in Fig. 3.59. Here, the same rice bran oil treatment was used. The correlation function is given by the following empirical equation:

$$\Delta L^*/L^* = 0.2186(\Delta R_m/R_m)^{1.3}$$

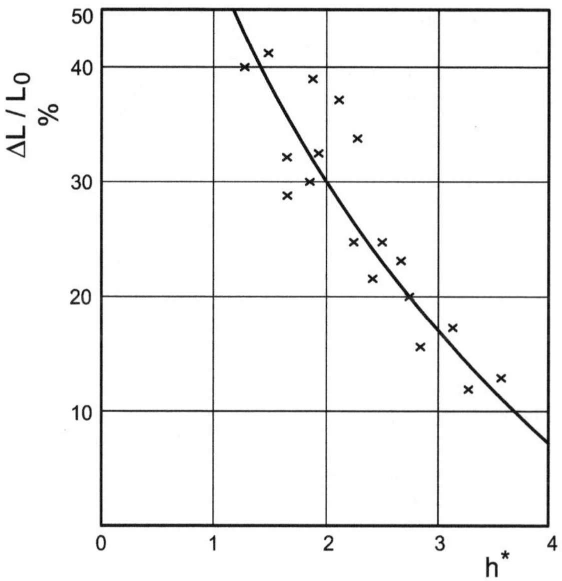

Fig. 3.58 Lightness reduction ratios as a function of colour hue due to rice bran oil treatment (Malaysian timber species)

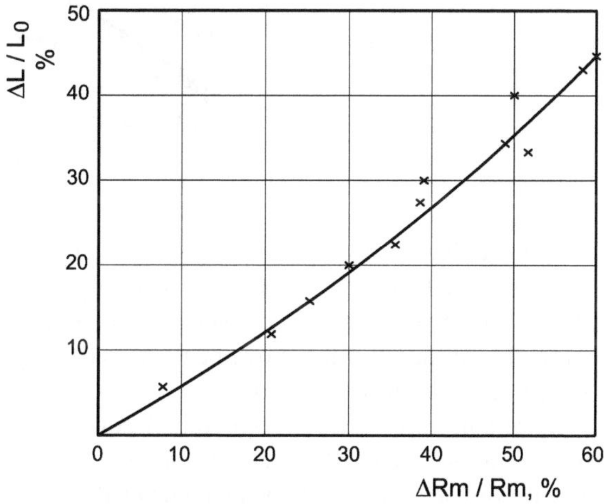

Fig. 3.59 Interrelation in relative changes of standardised lightness and physical lightness

The correlation given in Fig. 3.58 suggests that there may be a correlation between the lightness/saturation ratio L^*/C^* and the colour hue λ_0. This correlation is shown in Fig. 3.60 revealing a uniquely defined tendency although with some scattering of the measured data. This means that darker wood species with brown and reddish colours always have smaller lightness and a higher saturation of their colour. This finding is characteristic for natural wood materials and may not be valid for other materials.

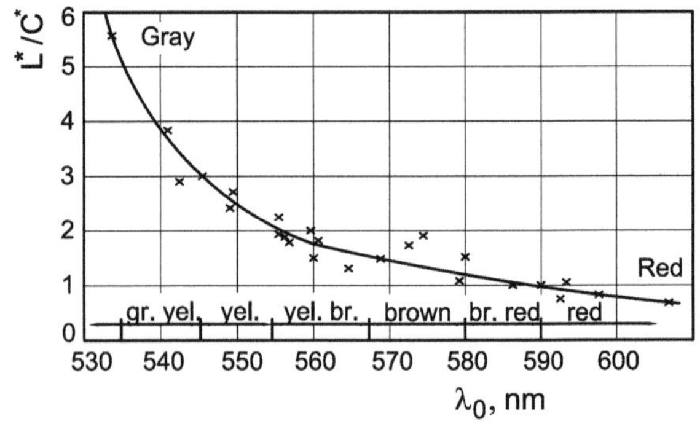

Fig. 3.60 Correlation between lightness/saturation ratios and colour hue for different wood species

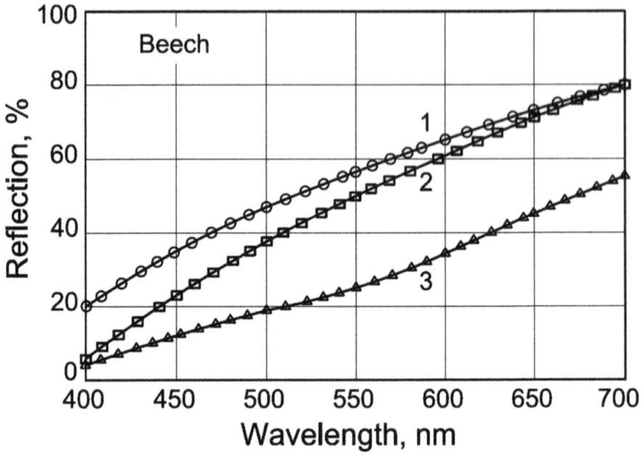

Fig. 3.61 Reflection spectrum of Beechwood. (*1*) Natural, (*2*) wetted, (*3*) coated with transparent lacquer

The wetting of wood surfaces has a similar effect to that of other colour enhancing methods. As the surface is drying out, however, the enhancing effect ceases. At the same time, the wetting brings about swelling and shrinkage just below the surface and the movements considerably modify the surface roughness (see more about in Sect. 4.10).

In certain cases, the clear-coating lacquer may interact with the timber and discolouration occurs. Figure 3.61 shows a similar case when the Beechwood coated with a transparent lacquer changed its colour to yellow. The reflection spectrum fundamentally differs from its original shape in natural condition. The distribution of reflection is almost linear characteristic for yellow colour.

3.8.3 *Representation of Colour Inhomogeneity*

Due to its cellular structure, wood never has a continuous colour distribution. The colouring materials, the extractives, are packed first of all in the cellular cavities of the heartwood. Making a cut through the timber, the cell wall shows its own colour, generally greyish yellow. Only the cavities packed with colouring materials show their characteristic colour.

Because of the small size of cell walls and cavities, our eyes see a more or less homogeneous colour from both the cell walls and cavities. Making a proper magnification of a cross section, the components of resultant colour can easily be distinguished. These inhomogeneities show no strikingly marked colour deviations and they may be termed as *microinhomogeneity*.

One of the most beautiful features of wood surfaces is their *macro-inhomogeneity,* diversity of which is almost infinitely high. Macro-colour inhomogeneities are mainly originated from the colour differences of earlywood and latewood, and the rays. Furthermore, irregular growth structures, infection with staining fungi, changes in the environmental conditions, the soil constituents and the age of the tree may also cause delicate differences in its local colour. In many cases, the lower part of the log, just over the root system, is very rich in irregular growth structures accompanied with vivid colour differences of various pattern.

Among the very diverse irregular growth structures, the big burls are particularly important. A round table made of one piece is unsurpassed in its striking beauty.

Finally, the log diameter and the direction of cut in relation to the longitudinal axis may results in countless diversity of colour pattern.

Our eyes perceive and judge a given colour mostly by its lightness and hue. Therefore, it is obvious to choose a similar coordinate system to represent colour variations for a given wood species. This representation does not give full picture of the beauty and aesthetical value of wood surfaces. Here, the colour pattern and its local distribution remain unknown. For a full visual observation, only a detailed photograph would be adequate. Nevertheless, this representation can give valuable information about the diversity of colour hues and their lightness.

The classification of colour patterns for wood surfaces is not an easy task and, to our best knowledge, such classifications did not come to light yet. A possible colour pattern classification is given below.

Colour pattern	
Fine scale	Large scale
Fine streaked	Blotched
Distinct vessels	Streaked
Distinct growth rings	Stripped
Fine dotted	Dotted
Irregular growth, fine	Curly
	Fiddle-back
	Bird's eye

The fine scale means sizes in the millimetre range while the large scale means sizes in the centimetre range. A fine scale colour pattern from a distance may seem to be a more uniform surface, because the densely packed fine streaks are apparently fusing together. Such wood species are, for example, Sucupira (*Andira micrantha*), Angelim pedra (*Hymenolobium modestum*), palm woods (*Cocos nucifera, Caryota urens*) and Rewarewa (*Knightia excels*) in tangential cut. Also distinct rings can create a stripped pattern when back sawn. The characteristic feature of a large-scale colour pattern is that the larger patches of different shape and colour appear on the base colour which is clearly visible also from a distance. Using more conventional wood species (Scots pine, Oak, Ash, etc.), large-scale colour pattern can be achieved by proper cutting directions, generally in tangential direction. Walnut species are richer in diversity of colour and, by proper selection and colour matching, they are suitable for high-quality furniture works.

It is clear that between the two pattern scales a continuous transition may exist, in many cases, as a function of cutting direction. Figure 3.62 shows fields of colour lightness and hue for different wood species (Sitkei 2013). Measurements were carried out on bigger surface areas and a small opening diameter (Φ 3 mm) was used. Latter is important in order to measure local values of lightness and hue.

The Ash (*F. excelsior*) has little variation in lightness. Its greyish yellow hue shows some variation from place to place. The two South American species, such as Angelim pedra and Sucupira, have considerable variation both in lightness and colour hue due to their fine streaked colour pattern. The rare Madagascan Bois de Rose (*Dalbergia maritima*), growing in the rainforest of north-eastern Madagascar, with its unrivalled lilac-red colour of different shades varies in lightness and hue.

The variation in lightness and hue has a general tendency: as the lightness increases the colour hue slightly shifts towards a greyish colour. It should also be noted that within the field of colour variation the change of colour hue is never continuous, which means, not all colour hues exist which could be present within the field. The extension of colour field is, however, very characteristic for a given wood species.

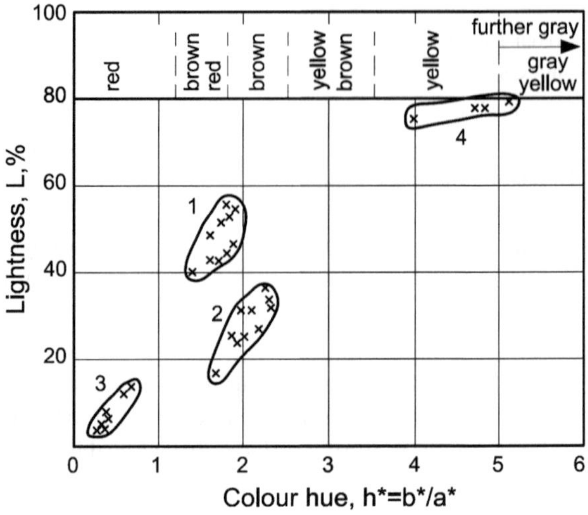

Fig. 3.62 Fields of colour variation for different wood species. (*1*) Angelim Pedra (*Hymenolobium modestum*), (*2*) Sucupira (*Andira micrantha*), (*3*) Bois de Rose (*Dalbergia maritima*), (*4*) Ash (*Fraxinus excelsior*)

In various wood species, the colour of latewood shows a distinct deviation from that of the earlywood (prominent growth rings). As an example, Fig. 3.63 depicts the colour fields for Scots pine with quite different colour hues. Both colour hues are almost uniform having a small extension of their field. The same figure shows the Black-heart Sassafras (*A. moschatum*) with a very wide range of colour characteristics. The colour hues extend in steps from the yellowish to the dark reddish. The two Peltogyne species from Central America have almost the same reddish colour hue with slightly different lightness. The Beefwood (*G. striata*) has darker coloured rays and an appropriate direction of cut reveals its characteristic meshed colour pattern.

In many wood species, the heartwood and sapwood have different colours. Sapwood generally has a light yellowish hue, while the heartwood depending on its extractives can show an unbelievably great variety of colours.

A special feature of the St. John's bread (*Ceratonia siliqua*) that it may have a wide range of colour hues in the heartwood as it depicted in Fig. 3.64. The correlation between lightness and hue follows the same tendency shown already in the previous figures. The Oregon myrtle (*Umbellularia californica*) with its yellowish brown colour has considerable variation both in lightness and hue.

Finally, Fig. 3.65 represents the colour fields for Kauri wood in natural and polished conditions, and swamp Kauri after 40,000 years. The colour hue varies much the same in all three cases but the lightness shows definite differences. With decreasing lightness, the colour saturation synchronously increases in such a manner that the lightness–saturation ratio slightly decreases compared to the machined surface.

The introduced new representation method is suitable to present and evaluate colour inhomogeneities, although not in all aspects. Further information can be obtained from colour pattern classification data and detailed colour photographs

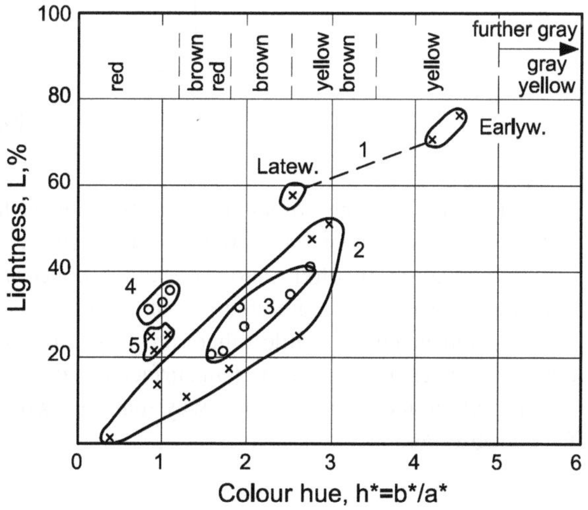

Fig. 3.63 Colour fields for different wood species. (*1*) Scots pine, earlywood and latewood, (*2*) Blackheart Sassafras (*Atherosperma moschatum*), (*3*) Beefwood (*Grevillea striata*), (*4*) Pau roxo (*Peltogyne excelsa*), (*5*) Nazareno (*Peltogyne purpurea*)

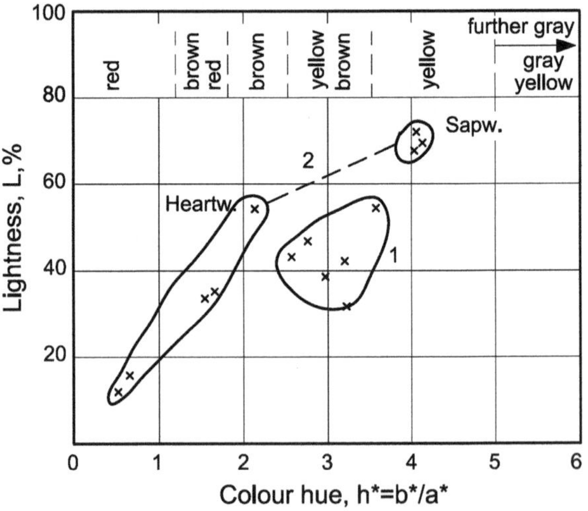

Fig. 3.64 Colour field variations for different wood species. (*1*) Oregon Myrtle, polished, (*2*) St. John's bread, polished

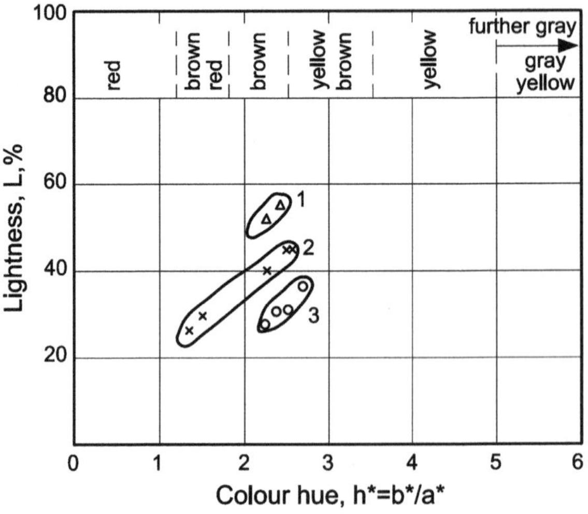

Fig. 3.65 Colour fields for Kauri wood. (*1*) Machined surface, (*2*) polished surface, (*3*) swamp Kauri, polished

(see Appendix II). The colour hue ranges given in the above figures facilitate to follow the actual colour hue variations and their extent.

The evaluation of colour inhomogeneities, concerning their aesthetical beauty, is a matter of taste. Some combinations of colour pattern and hues may be extremely pleasing and beautiful, while others are less attractive or even dislike. The art of design of an aesthetically important surface is hidden in the selection or matching of colour pattern and hues in such a way that they would be pleasing and expressive for the observer.

3.8.4 Colour Modification by Heat Treatment

Heat treatment is a common practice to equalise, stabilize or modify the colour of wood materials. Heat treatment can be accomplished by using heated dry or humid air, or by steaming which may be saturated or overheated. During high temperature treatment, the wood material may undergo browning reactions implying colour variation. The colour variation is a result of chemical reactions depending on temperature, duration of exposure and on the chemical constituents of the material. It is thought that the colour change is related to the change of double bounds in chromo-proteids (cromophore groups). Such double bonds can be found, for example, in lignin and extractives. Therefore, colour changes may mostly be attributed to the alterations in the extractives.

From early experiments (Stamm 1956), it is well known that heat treatment is followed by first-order reactions and the thermal degradation is accelerated in the presence of air and water. As a consequence, there are fundamental differences using heated dry air, humid air or steam.

The effect of heat treatment on colour change is fundamentally different from that of light (sun) irradiation (photodegradation). These two processes may interact and many several experimental works have combined heat treatment and light irradiation (Mitsui et al. 2001, 2004).

Figure 3.66 shows reflection spectra of Black locust wood in its natural condition (untreated), treated with dry air at 90 °C for 36 days and with steam at 90 °C for 36 days. The lightness and colour hue are also given for each case. The long-time treatment has made fundamental and different changes in lightness and colour hue (Tolvaj et al. 2010a).

It is very important to note that the heat treatment shifts the colour hue towards brown and red, while the lightness will be reduced considerably. It is remarkable, furthermore, that the presence of water (humid air) further increases the colour change and lightness reduction compared to the dry air treatment.

The effect of wetting without or with steaming is interesting. In Fig. 3.67, the reflection spectra of Beechwood are depicted for different treatments (Tolvaj et al. 2009). A simple wetting has similar effects to colour enhancement treatments with

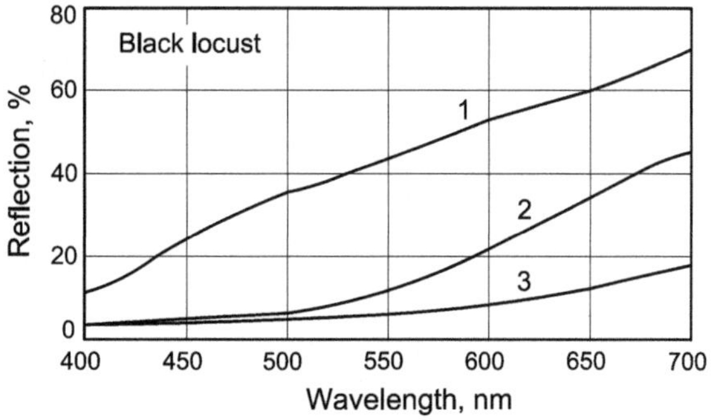

Fig. 3.66 Reflection of Black locust untreated (*1*), treated with dry air (2) and steamed (*3*)

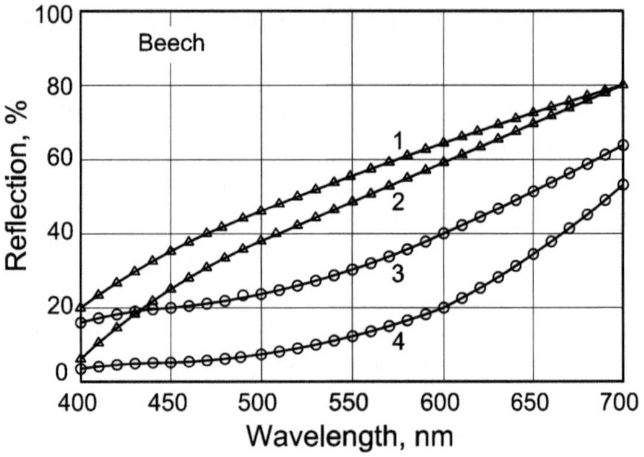

Fig. 3.67 Reflection of Beechwood under different conditions. (*1*) Untreated, (*2*) wetted, (*3*) steamed, (*4*) steamed and wetted

oil or wax. The reflection is mainly reduced in the 400–500 nm wavelength range and the colour hue changes very little. The steaming alone changes the colour hue considerably and the course of the reflection curve also differs from that of the untreated wood. An additional wetting further forces the effect of steaming and reduces the reflection almost evenly in the entire spectrum but the course of the reflection curve does not change much.

The considerable enhancing effect of wetting may be explained in the following manner. The saturated steam at 100 °C is in equilibrium with the wood surface at approximately 21 % moisture content. An additional wetting on the surface creates much higher moisture content considerably enhancing the hydrolysis.

As outlined above, the heat treatment is always accompanied with a colour change which is an important aesthetical factor. Despite this, no appropriate representation method has been developed to follow colour changes as a function of influencing factors such as temperature, the duration of heat treatment, steaming or irradiation.

It is a well-known practical observation that outdoor wood surfaces by the time will become more greyish over. However, in our climate, wood surfaces may be exposed to high temperatures up to 60–70 °C in a hot summer. In this case, the sun irradiation much more corresponds to a heat treatment, and as a consequence, the change of colour must be shifted also in this case towards brown and red and not towards grey! This was confirmed experimentally (Tolvaj and Mitsui 2010). Figure 3.68 shows experimental results for four wood species exposed to sun shine in a hot summer and the surface temperature could have been around 60 °C. The measurement data have been replotted representing the colour hue as a function of irradiation time. It can be stated that the colour hue is uniquely shifted towards the brown in all cases. The curves can be approximated with an exponential function in the form:

$$h^* = h_0^* - A\left(1 - e^{-t/T}\right) \tag{3.9}$$

where h_0* is the initial colour hue and A is the maximum variation of colour hue for a given treatment. The time constant T characterises the slope of the curves

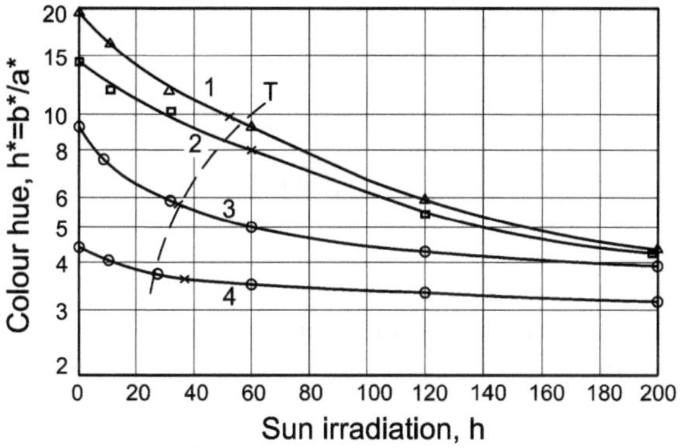

Fig. 3.68 Sun irradiation of different wood species in the hot summer with high surface temperatures. (*1*) Spruce (*Picea abies*), (*2*) Sugi (*Cryptomeria japonica*), (*3*) Black locust (*Robinia pseudoacacia*), (*4*) Beech (*Fagus crenata*)

(similarly to the relaxation time in rheology). As a rule of thumb, with increasing span of variation (A), the time constant T also slightly increases.

The temperature of a wood surface exposed to hot sun irradiation can be approximately estimated in the following manner. The thermal equilibrium of a surface depends on the heat flow from sun irradiation, on the heat conduction into the body and on the heat loss by convection into the air. The equilibrium of heat flows is described by the following equation:

$$I_{si} = I_{rs} + I_c + I_{co}$$

where

I_{si} is the total sun irradiation,
I_{rs} is the part of irradiation reflected,
I_c is the heat loss by convection, and
I_{co} is the heat conducted into the inner part of the body.

The heat convection into the air is given by the equation

$$I_c = \alpha(\vartheta_s - \vartheta_k)$$

where

α is the heat transfer coefficient,
ϑ_s is the surface temperature, and
ϑ_k is the air temperature.

Taking the wood body as the infinite half space, the heat flow can be calculated as (Carlslaw and Jaeger 1959) follows:

$$I_{co} = \frac{\vartheta_s \cdot \lambda}{\sqrt{4at/\pi}}$$

where
λ is the heat conduction coefficient of the wood,
a is the temperature conduction coefficient, and
t is the elapsed time.

Inserting the above two equations into the equilibrium equation yields

$$\vartheta_s = \frac{I_{si} - I_{rs} + \alpha \cdot \vartheta_k}{\alpha + \lambda/\sqrt{4at/\pi}} \tag{3.10}$$

Concerning the summer climate in Middle Europe, the following characteristic values may be chosen: $I_{si} = 3200$ kJ/m^2 h, $I_{rs} = 0.1I_{si}$, $\alpha = 25$ kJ/m^2 h °C (no wind) $\lambda = 0.8$ kJ/m h °C, $a = 0.00055$ m^2/h and $\vartheta_k = 25$ °C. Using these values, the surface temperature after 1 h will be 63 °C and after 2 h 75 °C. An equilibrium will be attained as heat convection becomes effective on all sides of the body, increasing the heat loss into the air.

The above experimental results call our attention to estimate the surface temperature in all cases when a wood surface is exposed to light irradiation. The thermal condition of the surface due to light irradiation decisively determines the direction of change in colour hue.

3.8.5 *Colour Change with Steaming*

Steaming is a common practice to enhance or homogenise the colour or to stabilize wood distortions. The temperature of steaming is generally around 100 °C and saturated steam is used.

Some wood species, such as Black locust, sometimes require colour homogenisation due to their inhomogeneous natural colour which makes it difficult to selected pieces to be colour matched. In particular, products with large surfaces, such as flooring or furniture fronts, require a proper selection or homogenisation of the raw material having very heterogeneous, unmatched colours. In this case, the colour change itself is not proper, and the colour differences should be reduced. The optimum selection of steam temperature and steaming time needs special attention for this purpose (Tolvaj and Molnar 2006).

Steaming may be combined with other treatments. Recently, it was discovered that an UV treatment before steaming amplifies the darkening effect of steaming and can provide a better and faster colour homogenisation (Mitsui et al. 2004).

The colour change due to steaming is also related to the change of conjugated double bonds mainly found in lignin and extractives. In Black locust, tannin may participate in condensation reactions caused by hydrothermal treatment in the temperature range of 90–120 °C.

It is noted that the steamed specimens should subsequently be conditioned at room temperature. Thereafter, new surfaces will be cut for colour measurements.

In the literature, the steaming process is generally followed on the $b^* - a^*$ plane and the resulting curve resembles a horseshoe, Fig. 3.69 (Németh 1982; Tolvaj et al. 2010b). This representation does not let us to see any colour change. This problem may easily be solved in such manner that we draw in the constant lines of $h^* = b^*/a^*$ and the corresponding colour ranges. The colour change and discolouration can clearly be followed for different steaming times. In Fig. 3.69, Black

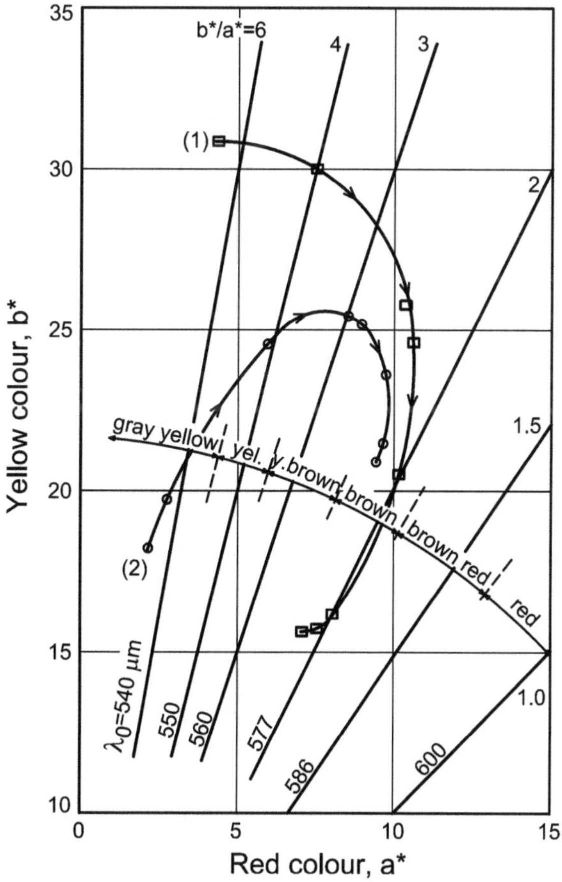

Fig. 3.69 Following the discoloration due to steaming on the $b^* - a^*$ plane. (*1*) Black locust (*Robinia pseudoacacia*), (*2*) Spruce (*Picea abies*)

locust (1) and Spruce (2) were steamed at a temperature of 90 °C and the colour changed from grey-yellow to brown. With the time, the change of colour is slowing down and after 10 days almost no change can be observed.

For practical uses, it is much better to present the measurement results, lightness and colour hue, as a function of steaming time at different steaming temperatures. From the course of the curves, the optimum process control can be established reading the time constant *T* for the given treatment.

Figure 3.70 shows the colour change of Black locust as a function of steaming time for different steam temperatures. It is general observation that the most intensive colour change occurs in the first 20 h. The steam temperature has a considerable influence on the colour change which can be characterised by the time constant *T* and used in Eq. (3.9).

As Fig. 3.70 shows, with increasing steam temperatures, the time constant rapidly decreases.

During the steaming process, the lightness of the timber decreases considerably following the exponential law with negative exponent. The variation of lightness for Spruce as a function of time is given in Fig. 3.71 (Tolvaj et al. 2012). Again, the

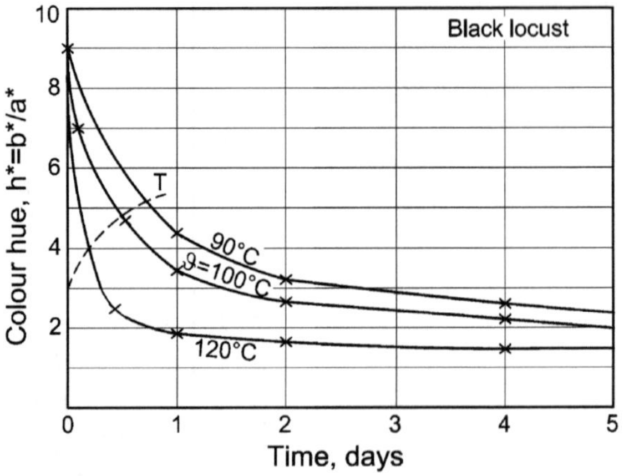

Fig. 3.70 The course of discolouration due to different steam temperatures as a function of time. Black locust (*Robinia pseudoacacia*)

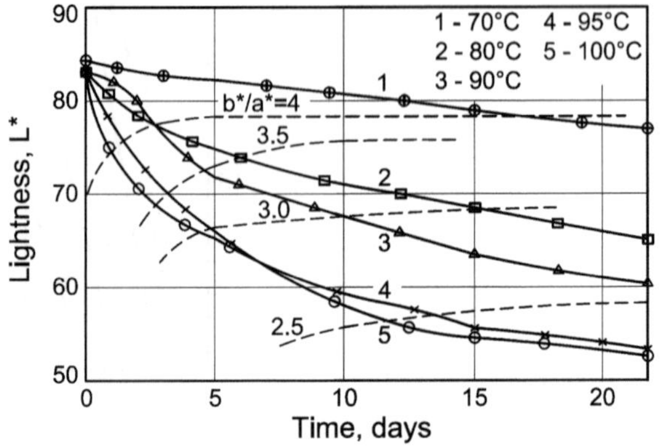

Fig. 3.71 Lightness versus steaming time for Spruce at different temperatures with constant colour hue lines

temperature has a fundamental influence and under 80 °C the lightness and colour hue change very slowly. In this figure, also the constant lines of $h^* = b^*/a^*$ are given and hereby the colour variations can easily be followed.

Concerning colour hue variations, the different wood species have similar behaviour although with some differences in their numerical values. The dynamics of colour changes for Spruce (more or less also for Fir) is reproduced in Fig. 3.72 (Tolvaj et al. 2012). It is again seen that the colour variation under 80 °C is very slow and the time constant T rapidly increases.

Some generalisation of experimental results, using a special representation form, seems to be possible. Using a temperature–time coordinate system, the lines represent constant lightness L^* or constant relative lightness reduction $\Delta L^*/L_0^*$ related to the initial lightness L_0^*. Such a representation for Black locust and Turkey oak is given in Fig. 3.73. The logarithmic scale facilitates a better read for small elapsed times. It is clearly seen again that the majority of changes occurs in the first 20 h of the treatment.

Some wood species have colour differences hindering their commercial utilisation. For example, Turkey oak (*Quercus cerris*) has a wide sapwood with light greyish colour impeding its use as a parquet flooring. Attempts were made to

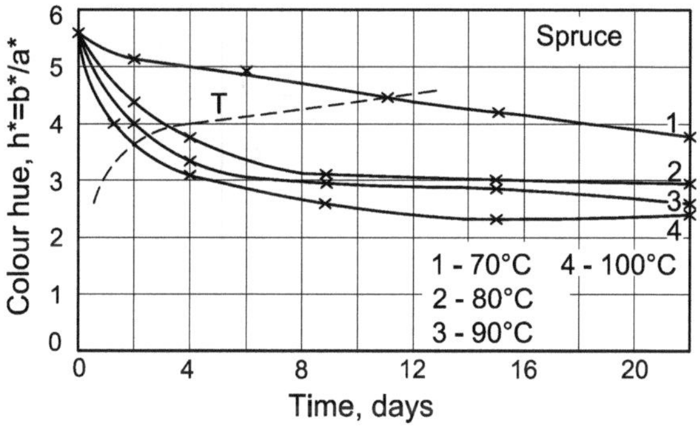

Fig. 3.72 Interrelation between colour hue and steaming time at different steam temperatures. Spruce (*Picea abies*)

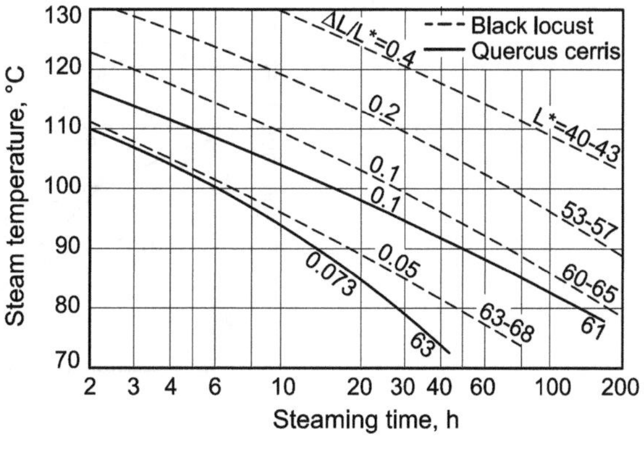

Fig. 3.73 Connection between steam temperature and steaming time for constant lightness reduction. Black locust, $L_0^* = 66 - 72$, *Quercus cerris*, $L_0^* = 68$

homogenise the colours of sapwood and heartwood by steaming. An important requirement was that the colour hues of both parts would be as near as possible, but the lightness cannot be very dark. Trials were conducted to establish proper steaming time which could fulfil the above requirements. Figure 3.74 shows the lightness variation as a function of steaming time for two steam temperatures (replotted from Tolvaj and Molnar 2006). The lightness of sapwood and heartwood was not far to each other after 6 h of steaming at 110 °C.

On the contrary, steam under 100 °C cannot diminish the difference in lightness in a proper extent, although a lightness reduction could have been achieved.

Concerning colour hue variations, the obtained results are depicted in Fig. 3.75 (replotted from Tolvaj and Molnar 2006). Again, using saturated steam, after 6 h the colour hues are very close to each other, while a steam under 100 °C can make

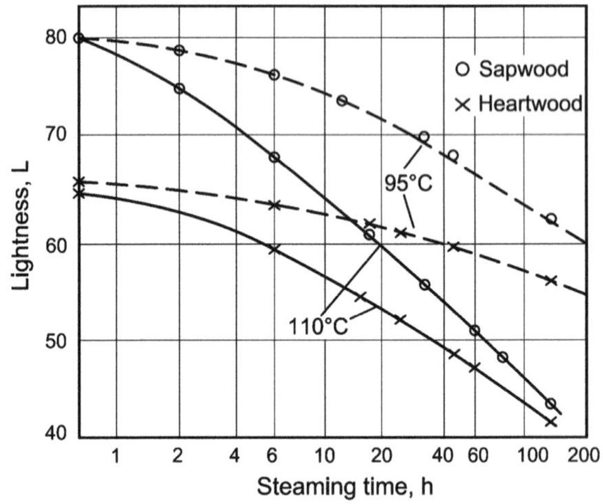

Fig. 3.74 The course of lightness reduction as a function of time at two steam temperatures

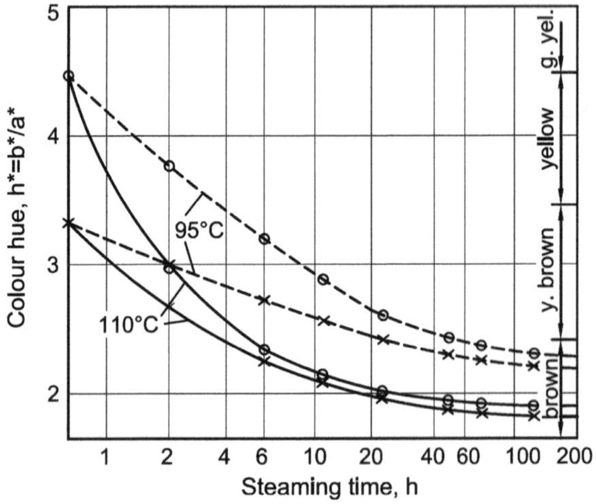

Fig. 3.75 Colour hue variation as a function of time for two steam temperatures. Turkey oak (*Quercus cerris*)

the hues similar only after 5–6 days. Therefore, a saturated steam of 110 °C and 4–6 h of treatment was recommended for colour homogenisation of sapwood and heartwood of Turkey oak.

For other wood species, different steaming parameters may be needed but, with all probability, they lie in this parameter range. It is most likely that steam temperatures under 100 °C are generally not suitable to homogenise wood samples of considerably different colours. In order to follow colour changes visually, the ranges of particular colour hues are also given in Fig. 3.75.

3.9 Colour Change Due to Sun Irradiation

Wood surfaces exposed to sunlight undergo photodegradation (see in Sect. 2.5) followed by structural deterioration and weight loss (Futó 1974, 1976). Due to the surface degradation, the reflection properties vary considerably, and therefore, the change in wood colour is the most sensitive indicator to demonstrate photodegradation. The comparatively high rate of colour change is generally attributed to the degradation of polyphenolic extractives present in wood and creating the particular colour of wood. Furthermore, also the lignin plays an important role in the photodegradation process.

Due to these observations, for modelling of photodegradation the light irradiation methods became popular using different light sources. Comparative studies revealed, however, that the induced photodegradation and colour change are highly dependent on the type of the light source (Tolvaj and Mitsui 2005). Because the UV component plays an important role in the photodegradation, the emission spectrum of the irradiator and with this the irradiated energy distribution as a function of wavelength are very important to know.

It should be stressed that there is a fundamental difference in colour change between the heating and light irradiation (without heat generation). At heat treatment, the colour hue shifts towards brown and red, while light irradiation moves it towards grey; (see Sect. 4.5). A combination of light irradiation and heat treatment may interact with each other and a previous light irradiation can enhance the effect of subsequent heat treatment (Mitsui et al. 2001, 2004).

Under Central European climate conditions, solar irradiation can be taken as "cool", i.e. without excessive warming-up of the wood surface, except for the hot months of July–August. Therefore, it is a common observation that outdoor wooden structures change their colour towards grey. Figure 3.76 shows experimental results on earlywood and latewood of Hinoki cypresse (*Chamaecyparis obtusa*) due to sun irradiation as a function of time (Tolvaj and Mitsui 2005). The earlywood has a smaller time constant T indicating that the earlywood is more sensitive to sun irradiation. The curves can be described with an equation similar to Eq. (3.9) given in Sect. 3.6 but changing the minus sign to positive. The h^*—values are increasing in both the cases indicating that the colour hue changes towards grey.

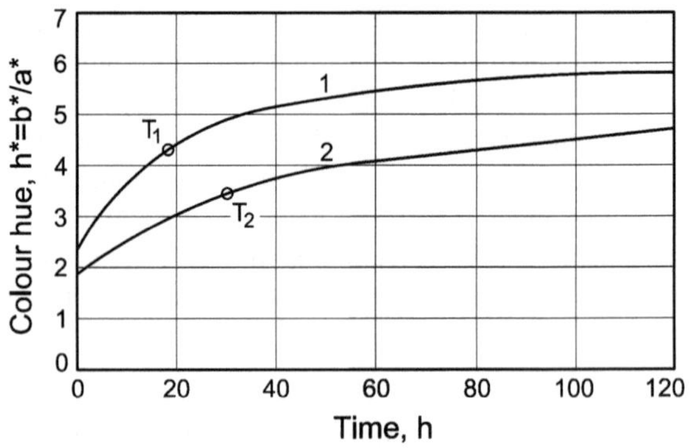

Fig. 3.76 Colour change of Hinoki cypresse due to sun irradiation. (*1*) earlywood, (*2*) latewood

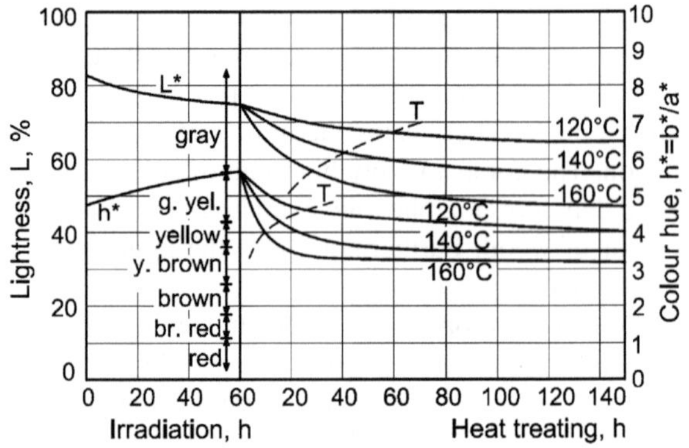

Fig. 3.77 Lightness and colour hue of Hinoki cypresse as a function of time with previous irradiation and subsequent heat treatment

A previous natural or artificial irradiation may change the dynamics of discolouration using subsequent heat treatment with dry or humid air. An interesting result is replotted from Mitsui et al. 2004 and depicted in Fig. 3.77. Hinoki samples were irradiated with an artificial light source (colour temperature 5500 K) having radiation heat flux of 2040 kJ/m^2 h for 60 h. Thereafter, a heat treatment was followed with dry air at different temperatures.

After irradiation, as it is expected, the lightness decreased and the colour hue changed towards grey. The thermal treatment further lowered the lightness, but the colour hue changed in the opposite direction, towards brown. The time constants *T* are also given for both set of curves. They are substantially different for lightness and colour hue.

The same picture is shown in Fig. 3.78 using humid air with 90 % RH, but with much lower temperatures. A comparison of the two figures reveals that the presence of water (humid air) highly accelerates the lightness reduction and colour change. It is turned out, furthermore, that the pretreating with irradiation, first of all, accelerates the lightness variation and influences the change in colour hue in a lesser extent. The saturation of colour, however, increases considerably due to the lightness reduction.

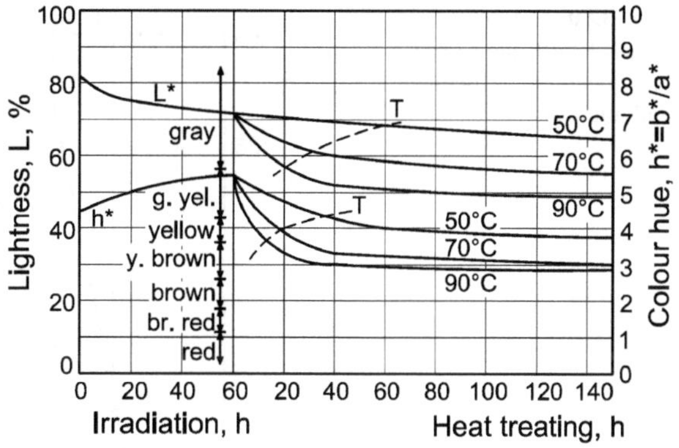

Fig. 3.78 Effect of heat treating with humid air on lightness and colour hue variation with preceding irradiation on Hinoki cypresse

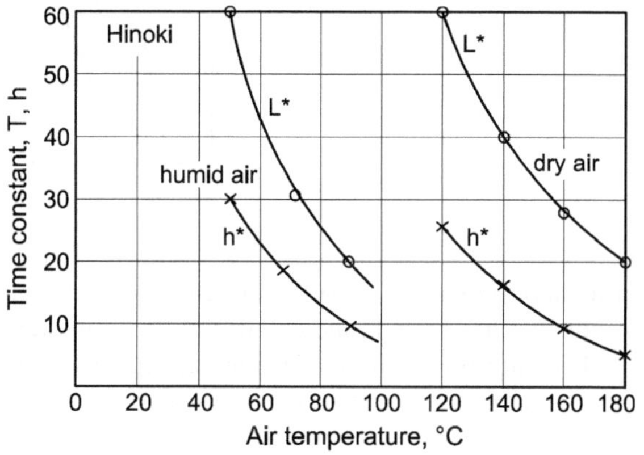

Fig. 3.79 Time constants for heat-treated Hinoki cypresse with preceding light irradiation

Based on Figs. 3.77 and 3.78, the variation of time constant T as a function of air temperature is depicted in Fig. 3.79. It can be stated that the use of humid air with RH = 90 % halves the required air temperature to achieve the same result either in light reduction or colour change compared to the dry air treatment. This finding is a newer proof to confirm the fundamental influence of water in the reaction mechanism of wood constituents especially extractives.

Another possibility for combined treatment is to apply a reverse sequence: first using a heat treatment and followed by irradiation (see also in Sect. 2.5.3). Black locust samples were heat treated with dry air of 160 and 200 °C temperatures up to 6 h which was followed by UV irradiation up to 36 h. Both treatment durations were varied. The direction of colour changes is plotted in Fig. 3.80. The initial greyish colour, due to the heat treatment at 160 °C, changed to brown and a subsequent UV irradiation produced practically no change in colour hue, although the colour saturation considerably increased. A more intensive heat treatment at 200 °C produced a brown-red colour, but this colour hue due to subsequent

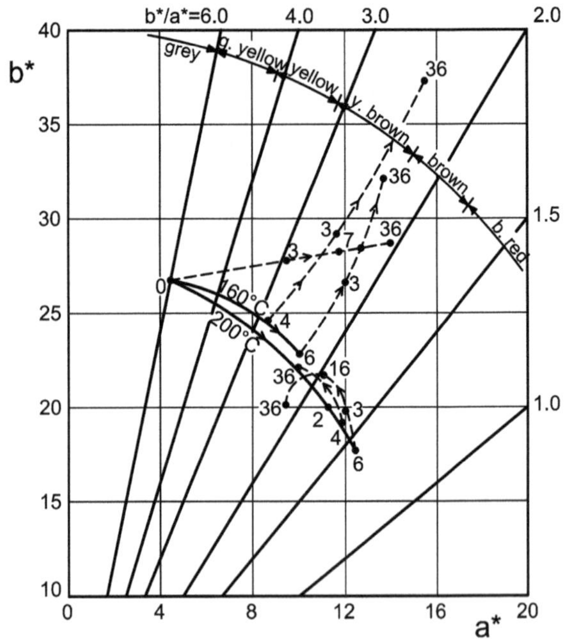

Fig. 3.80 Colour change of Black locust due to heat treatment and subsequent UV radiation

irradiation moved back to the brown range between b^*/a^* = 2.2 and 2.6. A pure irradiation has brought the colour into the same range.

The third possibility is the simultaneous irradiation and heat treatment (see also Sect. 2.6.2). An important difference is that now the UV irradiation and heat act simultaneously and strengthen each other (see Fig. 2.61). The measurement results for Pine and Ash are depicted in Fig. 3.81 using 120 and 160 °C air temperatures. The colour hue in all cases moves towards brown and its saturation rapidly grows to high values. Using 160 °C air temperature, the curves tend to take up the "horse-shoe" form (see Fig. 3.69).

With simultaneous actions, relatively low air temperatures (and surface temperatures) from 60 °C considerably strengthen the effect of UV irradiation and always result in more brownish colours. Even a high-powered light source can heat the surface fundamentally modifying the conditions.

Finally, it should be noted that existing experiments cannot answer for many practical questions. First of all quantitative relationships would be needed between irradiation intensity, time of exposure, wavelength distribution and the dynamics of discoloration for different colouring constituents. The accompanying surface photodegradation is the other side of this important question. Furthermore, a more rational processing and representation technique would facilitate the practical use of existing results to perform modelling for design purposes, including colour modifications, damage analysis and lifespan of outdoor structures.

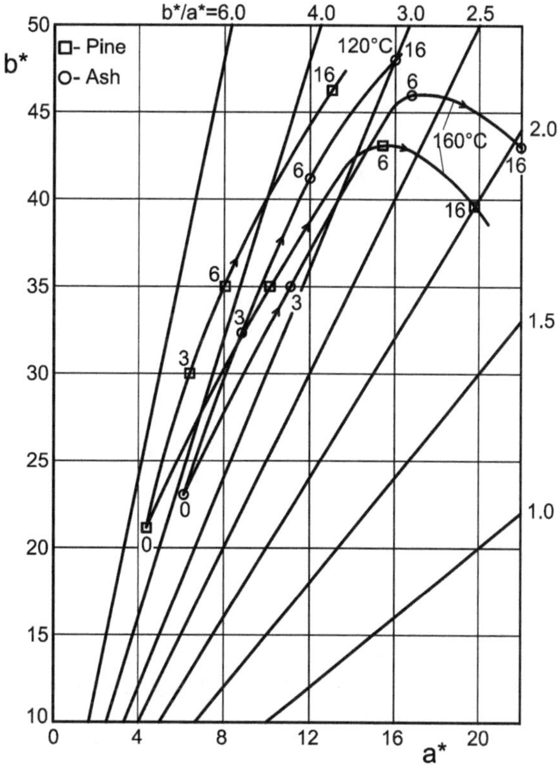

Fig. 3.81 Change of colour hues for Pine and Ash due to simultaneous heat and UV irradiation treatment. Numbers at the curves mean treatment duration in hours

3.9.1 Weathering of Wood and Its Modelling

Wood is a very versatile structural material having excellent physical and mechanical properties and it is widely used for outdoor applications. On many parts of the world, wood plays an important role in national cultures where households and religious items are made of wood. Many wooden structures, such as buildings, sculptures and monuments, withstood environmental actions for centuries and the oldest Japanese temples are well over one thousand years old. The strong affinity of people to the wood may be explained by its distinct beauty, its warm to touch, aesthetical appeal, low density, good thermal properties and, at the same time, excellent mechanical strength and that it is easy to work.

Wood is a biological material and, if it is not protected properly, is susceptible to various environmental effects such as solar radiation, moisture and temperature variations and air pollutants. Mainly UV light in sunlight is responsible for photochemical degradations. Due to the good light absorbing ability of wood, it interacts with different kinds of irradiations (UV, visible and IR) which leads to photoreactions causing photodegradation on the surface layer of the wood. As a consequence, the surface may undergo considerable changes in its general appearance suffering discoloration, loss of gloss and lowering other physical and mechanical properties. Among chemical constituents of wood, lignin and extractives play the most important role in the mechanism of photodegradation and subsequent discoloration.

Under the term *weathering*, we understand the single or combined effects of environmental actions taking place during longer time, generally years. The environmental actions may be quite different in their nature, e.g. their intensity, distribution in time and their combinations.

The main environmental actions are as follows:

- temperature, including very hot and freezing temperatures,
- humidity, including rain, snow and lasting high humidity,
- wind, enhancing the effects of temperature and humidity actions, and
- environmental pollutions, including acid rain.

A characteristic features of all environmental actions are their intensity and distribution in time, which make the problem very difficult to follow. These environmental loads on the wooden structures and on their surfaces are always unsteady causing permanent heat and moisture exchange accompanied by swelling and shrinkage. Due to the unsteady state condition, the gradient of heat and moisture distribution causes shrinkage stresses in the upper layer of wood and generally accompanied with permanent deformations modifying also surface roughness (see Sect. 4.10). The permanent movement of the surface facilitates the moisture infiltration further enhances the subsequent swelling and shrinkage.

It is clear that an accurate account of environmental actions is a very difficult task. For that reason experimental works usually only tell the measurement results without any deeper explanations. Standardised artificial weathering trials (cycles) allow researchers to more conveniently follow the entire cycle and make a more reliable evaluation and explanation. Therefore, for comparative studies, artificial weathering is more suitable. At the same time, the old problem is not solved: how to extend and use artificial weathering results for real outdoor conditions.

Figure 3.82 shows experimental results obtained in a two years long outdoor weathering trials using Black locust and Oak samples (Tolvaj and Papp 1999). This representation follows the colour change on the $b^* - a^*$ plane as a function of time.

The trial was started in a hot August, and therefore, in this month the heat effect prevailed. The greyish Black locust has quickly changed its colour towards brown as shown already in previous section. During the subsequent cooler season, the lightness and saturation of colour reduced continuously so that the colour hue slowly moved towards grey. The Oak and steamed Black locust originally had a yellow brown and brown hue, respectively, and during the hot period, first of all their lightness and saturation increased slightly. Thereafter, both samples had the same behaviour as the natural Black locust. After one year, most of the changes took place and after two years all samples had a quite similar greyish hue.

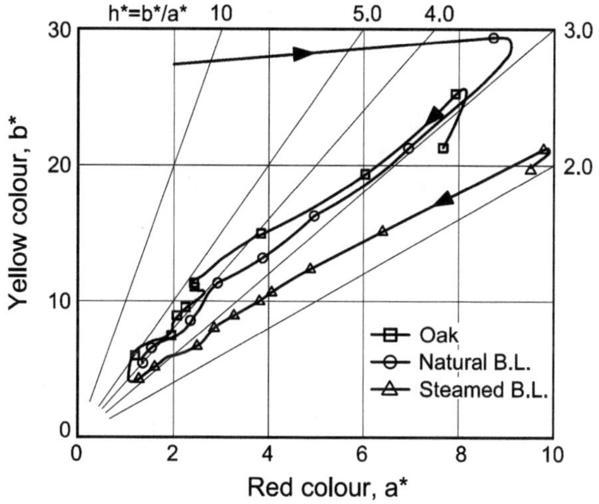

Fig. 3.82 Colour change of Oak and Black locust due to outdoor weathering during two years. The trials started in a hot August

The lightness variation of samples follows the weather changes in more explicit form. Figure 3.83 represents the lightness variation as a function of elapsed time. It is remarkable that the steamed Black locust, due to the hot August month, underwent a distinct lightening which increased the saturation mainly. During the cooler weather, all samples became darker but in the first 8 months they responded perceptibly to any changes in the course of weather. After one year also in the lightness almost no changes could have been observed.

As discussed earlier, the characteristic feature of the true grey colour is its constant reflection as a function of wavelength (Stefan 1879). Indeed, measurement of weathered wood samples gave reflection spectra close to the grey colour as it is shown in Fig. 3.84. After several years of weathering also the inner part of the samples was turning greyish to some extent.

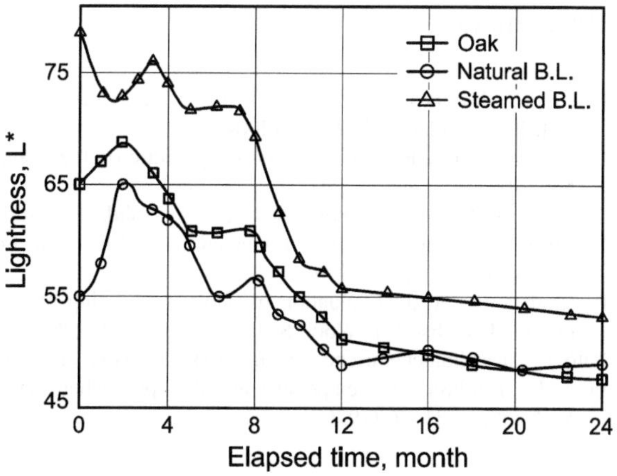

Fig. 3.83 Variation of lightness for Oak and Black locust samples due to outdoor weathering as a function of time

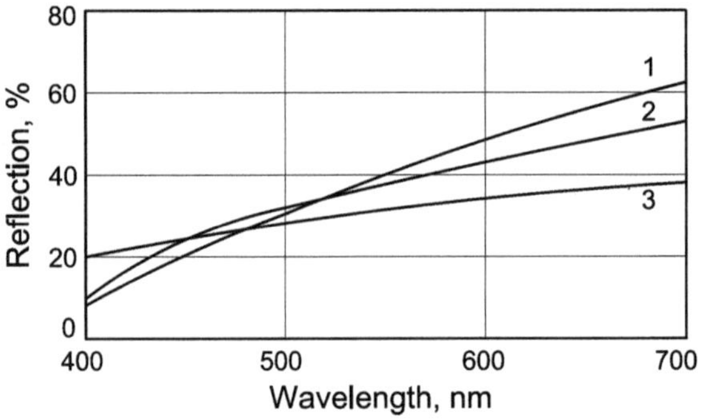

Fig. 3.84 Reflection spectra of a Spruce sample weathered over three years. (*1*) On new longitudinal cut, (*2*) on new butt-edge cut, and (*3*) on weathered surface

In order to protect wood surfaces under outdoor conditions, the use of different UV absorbers added to the coating material may be useful (Forsthuber et al. 2013; Pospisil and Nespurek 2000). These UV absorbers (such as benzotriazole) are effective by filtering out certain wavelengths of the light spectrum in the UV range. Mineral screening agents such as titanium dioxide, zinc oxide and iron oxide have been used for the same purpose. They have a strongly protective effect, probably because they prevent light penetration (Mills and White 2011; Forsthuber et al. 2013).

Standardised artificial weathering is frequently used for comparative studies. At the same time, up to the present no generally valid relationship has been established showing the effects of influencing factors such as temperature, moisture, irradiation and time. Therefore, the effect of real environmental conditions on the photodegradation and discolouration can be predicted very unreliably. An important task for future researches would be to pave the road in this direction.

References

Badami, S., Moorkoth, S., Suresh, B.: *Caesalpinia sappan*, a medicinal and dye yielding plant. Nat. Prod. Radiance **3**, 75–82 (2004)

Bekhta, P., Niemz, P.: Effect of high temperature on the change in colour. Holzforschung, 539–546 (2003)

Carlslaw, H., Jaeger, J.: Conduction of Heat in Solids. Oxford University Press, New York (1959)

Eckert, E.: Messung der Reflexion von Wärmestrahlen an technischen Oberflächen. Forschg. Ing. – Wes. pp. 265–270 (1936)

Forsthuber, B., et al.: Evaluation of photo- stabilizing efficiency of clear coatings comprising organic UV absorber. Wood Sci. Techn. 281–297 (2013)

Futó, L.: Der photochemische Abbau des Holzes. Holz als Roh- und Wekstoff, S. 303–311 (1974)

Futó, L.: Einfluss der Temperatur auf den photochemische Holzabbau. Holz als Roh- und Wekstoff, S. 31–36 (1976)

Masuda, M.: Why human loves wood grain figures. In: Proceedings of ICWSF, pp. 11–23. Ljubljana (2001)
Mills, J., White, R.: The Organic Chemistry of Museum Objects. Routledge, London (2011)
Minemura, N., Umehara, K.: In: Hokkaido Forest Products Research Institute Report, vol. 68, pp. 92 (1979)
Mitsui, K.: Changes in property of light-irradiated wood with heat treatment. Part 2. Effect of light-irradiation time and wavelength. Holz als Roh- und Werkstoff, 23–30 (2004)
Mitsui, K., Takada, H., Sugiyama, M., Hasegawa, R.: Changes in property of light-irradiated wood with heat treatment. Part 1. Effect of treatment conditions on the change in color. Holzforschung, 601–605 (2001)
Németh, K.: Evaluation of wood colours. Bull. Univ. Sopron, 1982 (in Hungarian)
Pospisil, J., Nespurek, S.: Photostabilization of coatings. Prog. Polym. Sci. 1261–1335 (2000)
Schmidt, E., Eckert, E.: Über die Richtungsverteilung der Wärmestrahlung von Oberflächen. Forschg. Ing. – Wes. pp. 175–183 (1935)
Sitkei, G.: Gloss of wood surfaces. Progress Report no. 2, DWE, University of West-Hungary (2014)
Sitkei, G.: Zur Bestimmung der Farbtöne von Holzarten. Fortschrittbericht. No. 2. DWE, University of West-Hungary Sopron (2013)
Stamm, A.J.: Thermal degradation of wood and cellulose. Ind. Eng. Chem. **48**(3), 413–417 (1956)
Tolvaj, L., Mitsui, K.: Light source dependence of photodegradation of wood. J. Wood Sci. **51**(5), 468–473 (2005)
Tolvaj, L., Mitsui, K.: Correlation between hue angle and lightness of light irradiated wood. Polymer Degr. Stab. 638–642 (2010)
Tolvaj, L., Molnar, S.: Colour homogenization of hardwood species by steaming. Acta Silv. Hung. **2**, 105–112 (2006)
Tolvaj, L., Papp. G.: Outdoor weathering of impregnated and steamed Black locust. In: Proceedings of ICWSF Conference Missenden Abbey, pp. 112–115 (1999)
Tolvaj, L., Nemeth, R., Varga, D., Molnar, S.: Colour homogenization of Beech wood by steam treatment. Drewno-Wood, 5–17 (2009)
Tolvaj, L., Takáts, P., Persze, L.: Effect of steam in the change of wood colour at 90 °C temperature. Faipar, 5–10 (2010a). (in Hungarian)
Tolvaj, L., Molnar, S., Nemeth, R., Varga, D.: Colour modification of Black locust depending on the steaming parameters. Wood Res. **55**(2), 81–88 (2010b)
Tolvaj, L., Papp, G., Varga, D., Lang E.: Effect of steaming on colour change of softwoods. BioResources **7**(3), 2799–2808 (2012)
Wyszecki, G., Stiles, W.: Colour Science. Wiley, New York (1969)

Chapter 4
Surface Roughness of Wood

4.1 Introduction

Roughness characterises the fine irregularities on a machined surface. These irregularities can be determined by measuring the height, width and shape of peaks and valleys produced by woodworking operations and anatomical structural properties. The measurement may be carried out in a plane (2D) or a given area (3D). Due to the complexity of the rough surface, a lot of roughness parameters were developed and proposed. There are international and national standards which define the accepted roughness parameters and give recommendations for their use.

The roughness of a wood surface depends on many influencing factors and it can be related to both wood properties and machining operational conditions. The important properties of wood are the tree species, density, moisture content and its anatomical properties. The internal structural properties include the specific number and the distribution of diameters of vessels and tracheids, and the portion of earlywood and latewood.

The machining process also has a significant effect on the surface roughness. The most important influencing factors are the cutting speed, tooth bite, the dullness of knife, the cutting angle, the direction of cut in relation to the grain and workpiece vibration.

Surface roughness is important in the general evaluation of the quality wood surfaces. The aesthetic value of a wood product is strongly related to its surface topography. Finished surfaces are highly sensitive to light reflection from any waviness, discontinuities and roughness differences on the surface. The gloss of a surface strongly depends on its roughness. The colour properties may also be modified by the roughness, first of all the lightness but much less the colour hue.

Surface roughness is important in wood technology. For instance, the surface stability and photodegradation are also influenced by surface roughness. The surface wettability and contact angle are also affected by the roughness which plays an

E. Csanády et al., *Quality of Machined Wood Surfaces*,
DOI 10.1007/978-3-319-22419-0_4

important role in the selection of an optimum machining operation prior to coating and staining.

One of the main difficulties in the research on wood surface roughness was the high variability in the internal structure of wood species, making it necessary to treat each wood species separately. This circumstance made it difficult to recognise generally valid relationships. In order to overcome these difficulties, a new *structure number* was introduced which uniquely characterises the internal structure of different wood species with respect to their expected surface roughness. The use of a structure number made it possible to establish some general regularity valid for all wood species.

The roughness component due to machining is partly related to the action of working edges which generally have cylindrical (knives) or spherical shapes (sanding). Using the laws of engineering mechanics, some interesting interrelations were derived and a comparison was made between the two edge shapes. Investigations have shown that the edge radius and the roughness parameter R_k are in strong correlation.

It has long known that the roughness measurements are always associated with relatively high scattering. The main reasons for the high standard deviation were established and related to various structural properties.

4.2 Surface Texture of Wood in Relation to Machining

The most important characteristic feature of wood is its porous structure and the presence of water in it. At the same time, wood is highly anisotropic and inhomogeneous. This is reflected in its physical properties and also in its surface texture due to machining. Furthermore, we can distinguish softwoods (conifers) with greater uniformity and hardwoods which are much more numerous, covering the tropical and temperate regions all over the world. Hardwoods have greater structural variation and they are much more variable in their physical properties, strength, density and also colour. Finally, both types of woods have earlywood and latewood with quite different density, void ratio and colour.

Wood has three major components: cellulose giving most of the mechanical strength, hemicellulose providing the matrix and lignin binding the cells together. Heartwood contains organic compounds and extractives, which may modify some physical properties, and they are responsible for the distinct colour of wood.

Depending on the environmental conditions, the thickness of annual rings may vary in a wide range, but the genetic properties of single species are also a definite influencing factor. There are very slow growing species (Huon pine, Kauri) and fast growing species (Radiata pine, young Redwood).

Figure 4.1 shows the microscopic photograph of a Scots pine. The principal portion of coniferous wood consists of longitudinal tracheids having thin walls in the earlywood and much thicker cell walls in the latewood. Accordingly, the inner diameters of lumens are considerably different in the earlywood and latewood.

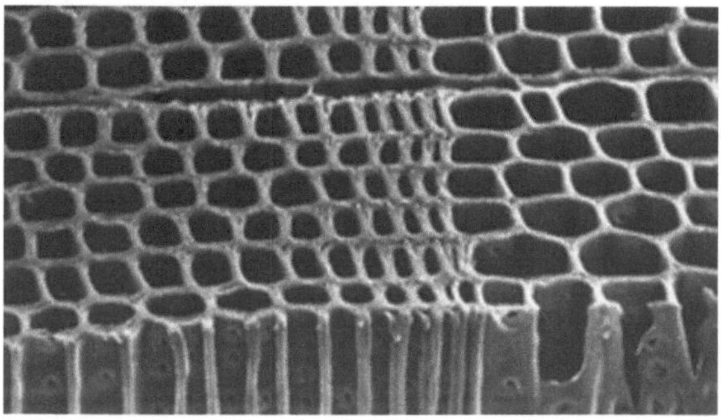

Fig. 4.1 Typical structure of conifers

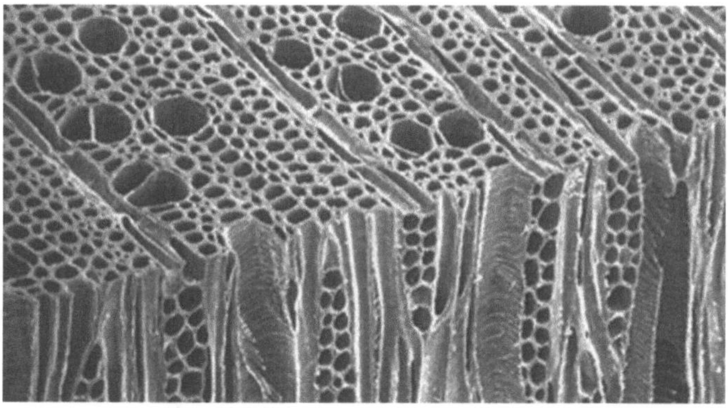

Fig. 4.2 Typical structure of hardwood

The relative portion of earlywood and latewood depends on the wood species and environmental conditions. As a rule of thumb, the thickness of earlywood is twice that of latewood.

A typical hardwood structure is shown in Fig. 4.2, which is much more complex compared to conifers. It consists of vessels, fibres and rays in both earlywood and latewood. The placement of vessels may be classified into diffuse porous and ring porous. The diameter of vessels, (depending on species and environmental conditions), may vary between 50 and 300 μm.

Particularly, Oak species, a typical ring-porous structure, have very large earlywood vessels with diameters up to 350 μm.

During machining, the cavities will be cut at different angles. Therefore, even with perfect cutting without any deformation of the surface, hollows do remain on the surface. These valleys cause a certain roughness which is not affected by the machining process. Therefore, the roughness evolving in this way is termed *structural* or *anatomical roughness*.

To measure roughness, the internal structure of wood can be characterised by the average diameter and number of cavities (vessels and tracheids) in the unit cross section. The size and number of cavities have to be determined both in the earlywood and in the latewood, and the ratios of earlywood to latewood should also be established.

The determination of these data is made on small-sized specimens taken from different wood species using a measuring microscope and a digital image processing technique. It is wise to check the results by calculation. Knowing the bulk density and the true density (1.5–1.52 g/cm^3) of the wood sample, the following approximation equation is valid (referred to the unit volume):

$$\rho_t = \left[\left(1 - \frac{d_1^2 \cdot \pi}{4} \cdot n_1\right) \cdot a + \left(1 - \frac{d_2^2 \cdot \pi}{4} \cdot n_2\right) \cdot b\right] \cdot \rho \tag{4.1}$$

where

ρ_t, ρ bulk density and real density of wood, respectively;
d_1, d_2 mean diameter of cavities in the earlywood and latewood;
n_1, n_2 number of cavities on the unit surface in the earlywood and latewood;
a, b earlywood and latewood portion ($a + b = 1$).

It should be stressed that the above equation is not very accurate; nevertheless, it is suitable to recognise greater measurement errors.

Table 4.1 summarises the typical characteristics of the anatomical structure of conifers and hardwood species tested (Magoss and Sitkei 2001).

Notations in Table 4.1: $\bar{d}_i$ is the average diameter of vessels and tracheids, $\bar{n}_i$ is the average number of vessels and tracheids in the unit cross section, $\bar{a}$ is the average portion of earlywood, and $\bar{b}$ is the average portion of latewood.

It is interesting to note that a good correlation exists between the diameter of vessels or tracheids and their specific number. This correlation for conifers and hardwood species is depicted in Fig. 4.3. The distribution is of parabolic type (straight line on double logarithmic scale), and conifers have a slightly higher tracheid diameter for the same specific number in the unit cross section. This diagram is also suitable to check new measurement data.

The cutting of vessels, tracheids and other void elements causes surface irregularities independently of the machining process itself. To characterise the anatomical roughness of a wood structure, the following consideration is taken into account (Fig. 4.4). Each cut vessel and tracheid contributes to the total roughness in a measure proportional to their cross sections in the unit length. A characteristic number can be found in the following form (Magoss and Sitkei 2001):

$$\Delta F = \frac{\pi}{8}\left[a \cdot \left(\sqrt{n_1} \cdot d_1^2 + \sqrt{n_2} \cdot d_2^2\right) + b \cdot \left(\sqrt{n_3} \cdot d_3^2 + \sqrt{n_4} \cdot d_4^2\right)\right] \quad [\mathrm{cm}^2/\mathrm{cm}] \tag{4.2}$$

Table 4.1 Structural property of the tested wood samples

Wood species	Earlywood			Latewood		
	$\overline{d_i}$ (μm)	$\overline{n_i}$ (piece/cm^2)	$\overline{a}$	$\overline{d_i}$ (μm)	$\overline{n_i}$ (piece/cm^2)	$\overline{b}$
Thuja	26.5	142,800	0.8482	14.0	316,600	0.1518
Spruce	30.0	111,335	0.8478	19.0	160,400	0.1522
Pine	28.0	125,100	0.6694	20.0	135,840	0.3306
Larch	38.0	65,490	0.6310	17.5	145,000	0.3690
Beech (vessel)	66.0	15,740	0.7000	48.0	14,020	0.3000
Beech (tracheid)	8.2	342,890		6.4	490,290	
Oak (vessel)	260.0	400	0.5900	50	12,000	0.4100
Oak (tracheid)	22.5	120,000		19.6	85,000	
Black locust (vessel)	230.0	546	0.5800	120.4	1,500	0.4200
Black locust (tracheid)	15.0	270,000		9.6	280,000	
Cottonwood (vessel)	69.7	9,500	0.6666	44.0	12,700	0.3333
Cottonwood (tracheid)	12.7	309,500		11.0	300,892	
Ash (vessel)	177.0	670	0.6100	52.00	7,500	0.3900
Ash (tracheid)	19.0	190,000		14.0	230,000	

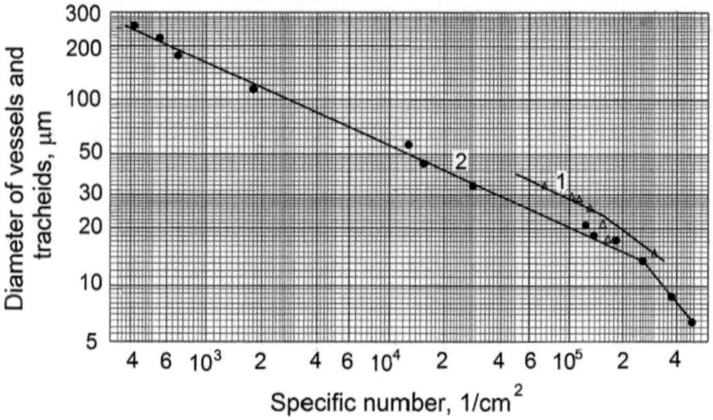

Fig. 4.3 Correlation between the diameter of vessels and tracheids and their specific number in the unit cross section. *1*—conifers; *2*—hardwoods

where

n_1, n_2 the number of vessels and tracheids in the earlywood in the unit cross section;

n_3, n_4 the number of vessels and tracheids in the latewood in the unit cross section;

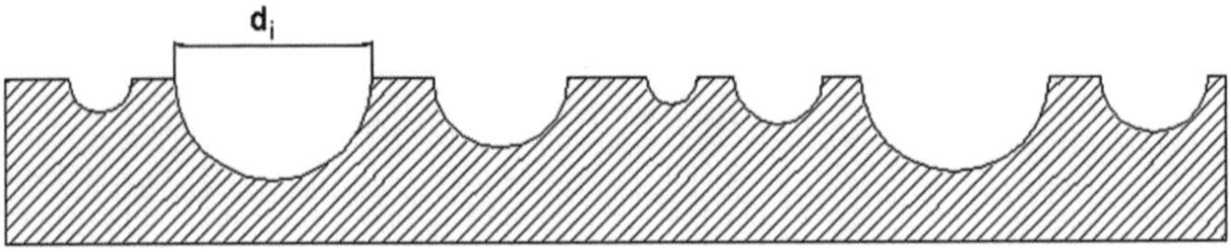

Fig. 4.4 The model of structural surface roughness

d_1–d_4 the diameter size of vessels and tracheids in the earlywood and latewood, respectively;

a, b portions of earlywood and latewood, ($a + b = 1$).

The value ΔF defined with the Eq. (4.2) is called *structure number*, which gives an accurate definition of each wood species based on the size and specific number of cavities in the wood structure. Accordingly, surface irregularities caused by the internal wood structure are expected to have a definite correlation with the structure number.

A further advantage of the structure number is that it enables the characterisation of wood species based on their internal structure and helps to establish correlations among the surface roughness parameters.

It is well known that results of surface roughness tests usually show significant scatterings. One of the reasons for this is the accidental position of tracheids and vessels on the machined surface. A further significant scattering can be caused by the accidental position and cut of the earlywood and latewood or the seasonal change of the earlywood/latewood ratio. Namely, the value ΔF may have substantial differences in the earlywood and latewood. Pine species show the smallest deviation, whereas hardwood species generate a significantly bigger one. Figure 4.5 illustrates the alteration of the ΔF value of different wood species depending on the earlywood ratio. The starting point of the curves (a = 0) indicates the pure

Fig. 4.5 Alteration of the ΔF value of different wood species in relation to the earlywood ratio

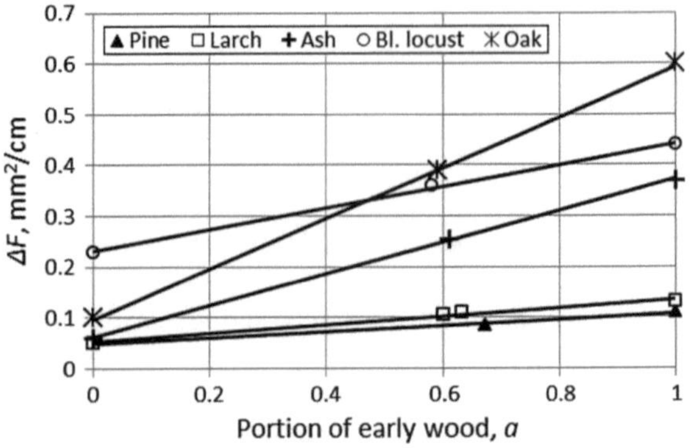

latewood, while the end point ($a = 1$) indicates the pure earlywood. Oak shows the biggest change, but Ash also shows a significant one as well.

Tests show that the earlywood ratio falls predominantly in the range $a = 0.5$–0.7 as shown in Fig. 4.5.

Therefore, it seemed practical to apply the relative changes of ΔF for the range $a = 0.5$–0.7 in accordance with the following equation:

$$\delta(\Delta F) = \frac{\Delta F_{0.7} - \Delta F_{0.5}}{\Delta F_{0.6}} \tag{4.3}$$

Wood species can be characterised with the ratio of cross sections of cavities cut in the earlywood and latewood, which we can express using the following Eq. (4.2):

$$B = \frac{d_1^2 \cdot \sqrt{n_1} + d_2^2 \cdot \sqrt{n_2}}{d_3^2 \cdot \sqrt{n_3} + d_4^2 \cdot \sqrt{n_4}} \tag{4.4}$$

Since Pine species have the vessels missing, the Eq. (4.4) for conifers will be simpler.

Figure 4.6 shows the relative changes of ΔF in relation to the parameter B defined with Eq. (4.4).

The apparent correlation between the two variables is displayed clearly. However, it is remarkable that beech is located right next to Scots pine, while Black locust shows a smaller relative change than Larch. Consequently, the relative change has no connection with the absolute value of ΔF.

The correlation obeys the following empiric equation:

$$\delta(\Delta F) = 7.8 \cdot B^{0.65} \quad [\%] \tag{4.5}$$

In conclusion, wood structure is a major influencing factor which fundamentally contributes to the roughness of wood surface.

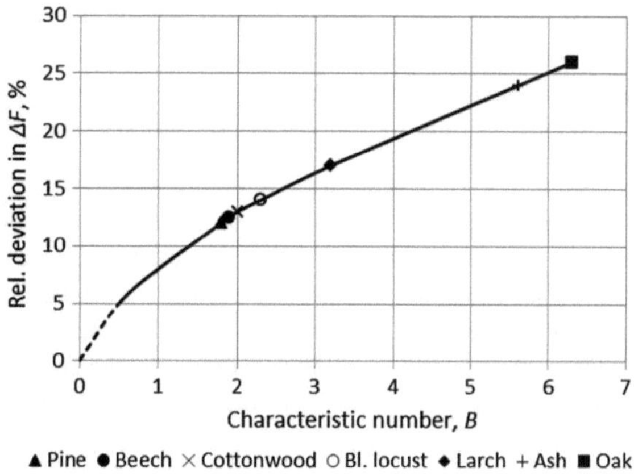

Fig. 4.6 Relative changes of ΔF in relation to the parameter B

4.3 Origin of Surface Roughness

Roughness or surface topography is a characteristic feature of machined surfaces. The separation of chips from the surface is always associated with force–deformation phenomena causing rupture in the separation zone (Csanády and Magoss 2013). Due to the finite radius of the knife, deformations occur under the separation plane which are partly elastic (with subsequent recovery) and partly permanent.

Wood is a capillary-porous material with a considerable void ratio. Depending on the local void ratio (earlywood or latewood) and on the edge radius of the knife, compaction of the material takes place in the thin upper layer. Compaction means that material is squeezed out from a given volume and pressed into the layer just below.

Woodcutting belongs to the group of so-called free cuttings. Its main characteristic feature is the absence of a counter-edge to ensure the required counterforce. In the process of free cutting, the counterforce is produced by the strength of material and by inertia forces. Therefore, woodcutting always requires high cutting speed.

Another inherent property of wood is its brittle fracture which is further complicated by the strong anisotropy of timber materials. All wood materials have low tensile strength perpendicular to the grain which may contribute to fibre separations and grain raise.

Due to its capillary-porous nature, wood has two major roughness components: one is due to machining and the other one is caused by the anatomical structure of wood (Magoss 2008, 2013). Even in the case of a perfect cutting, inner cavities will be cut, giving discontinuity in the surface which appears as roughness.

If we consider the three height components of roughness R_{pk}, R_k and R_{vk} (see in Sect. 4.4), then the following may be stated:

- The reduced peak height R_{pk} is fundamentally determined by the machining process for a given wood species. The mechanical properties of the timber, such as hardness (density) and the direction of cut, may have definite influences on different wood species.
- The core thickness R_k is influenced mostly by the sharpness of the tool edge, but the anatomy of wood also has some effects.
- The reduced valley height R_{vk} is fundamentally governed by the anatomy of wood and only in a lesser extent by the other influencing factors.

As Sect. 4.10 will describe in detail, an occasional wetting of the surface, due to swelling and shrinkage, will rearrange the initial values of R_{pk}, R_k and R_{vk}.

The tool edge separating chips from the surface imposes various loads on the contacting material. Compressive forces cause the compaction of the material with considerable crushing effects, while tensile forces may produce fibre separation and local rupture in the immediate vicinity of the moving edge. Therefore, it is interesting to examine the interaction of a rounded edge with its neighbourhood.

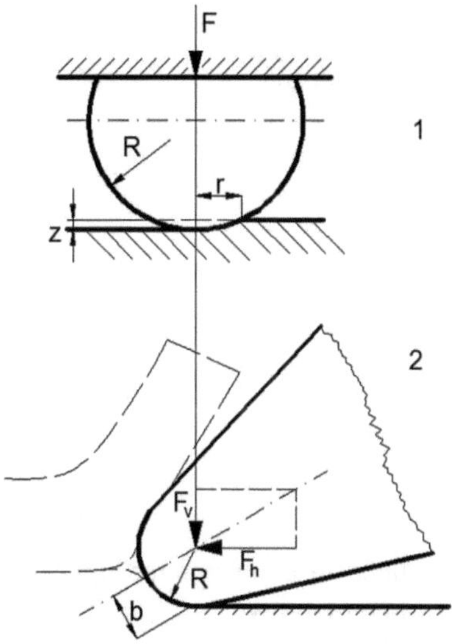

Fig. 4.7 Deformation under a spherical (*1*) and cylindrical body (*2*)

The most commonly used tool profiles are a knife-edge and spherical grit for sanding, as shown in Fig. 4.7. The mechanics of an infinite half-space means that (Timoshenko 1951), a knife-edge corresponds to an infinitely long cylindrical body contacting the plane. Spherical grit corresponds to a spherical body contacting the plane surface. In the following, we examine the stress and deformation under these cutting edges with special attention to their effect on the R_k layer.

The radius of new or sharpened tool edges, after a short running-in period, amounts to around 15–20 μm. Due to the wear process, the radius of a tool edge continuously increases. The edge radius between 20 and 40 μm, in its "working sharp" stage, ensures good cutting and good surface quality. Over an edge radius of 40 μm, in the blunting stage, the tool works with increasing cutting forces and decreasing surface quality (Csanády and Magoss 2013).

The grit diameter is determined by the number of meshes per inch of the sieves used for screening. The real size of the grits is generally taken as normally distributed between two successive sieves. In practice, the average grit size diameter is used, as given in Fig. 4.8, as a function of standard grit size notation.

The grits are embedded in the base carrying material in a very great numbers, and therefore, they overlap each other many times in their pathway. The most important operational parameters are the surface pressure, the feed speed, the grit size and the cutting speed. Several other influencing parameters may be mentioned such as the hardness of the abrasive material, its heat conduction coefficient and the contact lengths of the sanding .

In all woodworking operations, the spherical edge of a knife exerts a stress field in the underlying material causing distortions and ruptures. This stress field may considerably influence the resultant surface roughness depending on the edge geometry and material properties. A negative rake angle of the cutting edge is

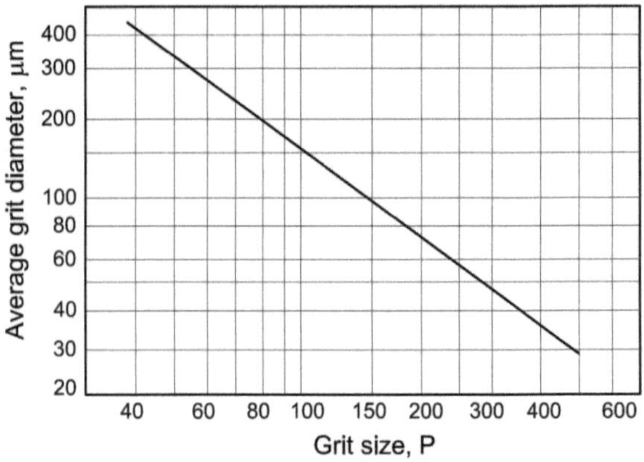

Fig. 4.8 Average grit diameter as a function of standard grit size

inclined to compress the upper layer of the surface which can be detrimental to surface stability.

The contact and the related deformations of a cylinder on a plane surface can be calculated according to the Boussinesq equation in the following manner. Taking the edge as a rigid body, the half-contact width is (Timoshenko and Goodier 1951):

$$b = \sqrt{\frac{4P' \cdot (1 - v)^2 \cdot R}{\pi \cdot E}} = 1.06 \cdot \sqrt{\frac{P' \cdot R}{E}}$$

The maximum deformation is given by

$$z = 0.567 \cdot \frac{P'}{E}$$

The maximum pressure in the centre line is

$$\sigma_0 = 0.6 \cdot \sqrt{\frac{P' \cdot E}{R}} \quad \text{and} \quad \frac{z}{b} = 0.88 \cdot \frac{\sigma_0}{E} \tag{4.6}$$

where

P' is the load per unit length on the edge, N/cm;
E is the modulus of elasticity of the timber, N/cm^2;
R is the edge radius, cm;
v is the Poisson's ratio, taken as 0.35.

Similarly, the contact of a spherical body on the plane surface gives the following expressions:

$$r = 0.87 \cdot \left(\frac{P \cdot R}{E}\right)^{1/3}$$

$$z = 0.7566 \cdot \left(\frac{P^2}{R \cdot E^2}\right)^{1/3}$$

$$\sigma_0 = 0.63 \cdot \left(\frac{P \cdot E^2}{R^2}\right)^{1/3} \quad \text{and} \quad \frac{z}{r} = 1.38\frac{\sigma_0}{E} \tag{4.7}$$

where P is the load on the sphere, N.

Comparison of Eqs. (4.6) and (4.7) suggests that for the same maximum stress, a spherical body has a larger deformation than a cylindrical body.

An important practical question is the variation of maximum stress as a function of grit size, surface pressure and material properties.

Taking 1 cm^2 surface area, the number of particles depends on the particle diameter and the cover ratio:

$$n \cdot \frac{d_i^2 \cdot \pi}{4} = 1.0 \cdot \Psi \quad \text{and} \quad n = \frac{1.273 \cdot \Psi}{d_i^2}$$

where

d_i is the particle diameter, cm;

Ψ is the particle cover ratio on the surface.

Exerting a p pressure onto the surface (N/cm^2), the force acting on a single particle is calculated as

$$P_i = \frac{p}{n} = \frac{p \cdot d_i^2}{1.273 \cdot \Psi}$$

The radius of deformation is given by the known Boussinesq equation and making use of the previous equation:

$$r = \left(\frac{3}{4} \cdot P \cdot R \frac{(1 - \nu^2)}{E}\right)^{1/3} = 0.637 \cdot d_i \cdot \left(\frac{p}{\Psi \cdot E}\right)^{1/3}$$

and the maximum contact pressure under the grit is

$$\sigma_0 = 0.923 \cdot \left(\frac{p}{\Psi}\right)^{1/3} \cdot E^{2/3} \; [\text{N/cm}^2] \tag{4.8}$$

where E is the modulus of elasticity of the wood material, N/cm^2.

Equation (4.8) clearly shows that the maximum stress under the grit does not depend on the grit size.

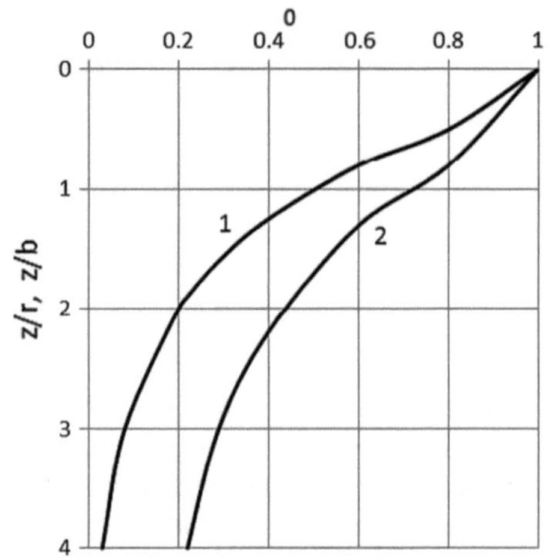

Fig. 4.9 Stress relations under a spherical (*1*) and cylindrical body (2) using hemispherical pressure distribution in the contact surface

The above considerations and calculations refer to a solid body without cavities. Therefore, depending on the internal structure of wood, much higher local stresses may be expected which will be capable of rupturing the material. Due to the horizontal movement of the edge, a shear stress field will be superimposed which further enhances the destructive effect of the vertical stress field.

The pressure distribution in the half-space under the contact surface is shown in Fig. 4.9. Under a spherical body, the decrease of normal stresses occurs more rapidly compared to those of a cylindrical body as a function of the relative depth. Due to the relative coordinate (z/b, z/r), the absolute value of deformations increases in the same ratio as the radius or half-width of the loaded surface. Using a coarser grit size and the same overall surface pressure, the same stress level occurs deeper under the loaded surface, and therefore, the crushing effect of the vertical stress will be larger. The crushing effect of the vertical stress will be considerably enhanced by shear stresses appearing due to the horizontal motion of the working edges.

For a given case, the stress decay as a function of depth can easily be calculated. For comparison, Fig. 4.10 shows the rapid decrease of stresses under the surface for two sandpaper grades. It is clearly seen that the coarser grit exerts a given stress

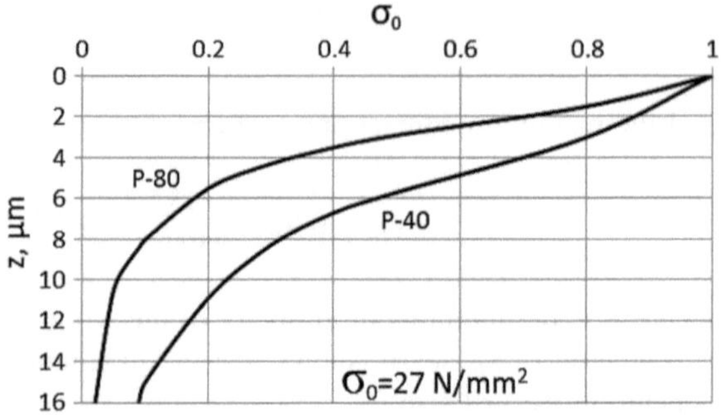

Fig. 4.10 Stress decay under grit as a function of depth. $E = 1400\ \text{N/mm}^2$, $p = 1\ \text{N/cm}^2$, $\Psi = 0.8$

roughly two times deeper than the finer grit. As a consequence, the crushing effect that corresponds to a given stress level also affects a deeper zone considerably increasing the R_k layer.

Keeping in mind the stress distribution in Fig. 4.9, it is obvious that a bigger radius of a sphere is equivalent to smaller radius for cylindrical edges concerning their effect on the R_k layer thickness. Experimental evidence supporting these considerations will be given in Sect. 4.8.

4.4 Roughness Characterisation

The surface topography of engineered surfaces is very complex, and in wood, it is influenced by countless factors. Some of these influencing factors relate to the machining process, while others relate to the anatomy of wood in question.

Due to the extreme complexity of the surface topography, it cannot be described by a few parameters. Each parameter can only characterise one aspect of topography. Therefore, in the history of surface roughness measurement (since the 1930s), a myriad of roughness parameters were proposed and many of them were introduced as international and national standards.

Measurement has several problems with consequences: the accuracy may depend on the filter cut-offs and on the sampling intervals. In the history of development, first a two-dimensional measuring and evaluation system was evolved mainly using stylus instruments. The stylus always has definite geometrical dimensions such as tip radius and cone angle which may be a limiting factor in detecting fine details. This effect is known as mechanical filtering.

The most commonly used roughness parameters are as follows:

Parameter	Symbol
Average peak-to-valley height	R_z
Maximum peak-to-valley height	R_{max}
Average (integrated) roughness	R_a
Root mean square (RMS) roughness	R_q
Skewness (the third moment of profile amplitudes)	R_{sk}
Kurtosis (the fourth moment of profile amplitudes)	R_{ku}

Figure 4.11 shows and explains the possible variations of skewness and kurtosis.

Since 1933, the Abbott–Firestone parameters have been in use, characterising the material and void distribution in rough surface. These parameters are the following (Abbott and Firestone 1933):

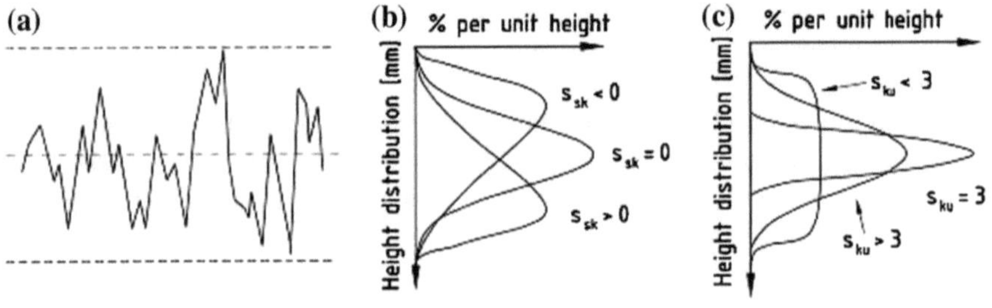

Fig. 4.11 Height distributions (**a**) for skewness (**b**) and kurtosis (**c**)

Parameter	Symbol
The average height of protruding peaks	R_{pk}
Vertical difference in the core section	R_k
The averaging height of valleys	R_{vk}
Material volume above the upper core surface	M_{r1}
Void volume under the lower core surface	M_{r2}

Figure 4.12 shows the Abbott–Firestone curve and the meaning of particular parameters.

Since the 1980s, an increasing need has arisen for characterisation of surface topography in three dimensions (3D). Some of the 3D parameters are extended from the previously defined 2D counterparts; others are newly defined for a more detailed characterisation of 3D topography. This proposed parameter set was determined to meet the demands of the metal industry (Dong et al. 1994).

The demands of the wood industry in many cases differ from those of the metal industry. In wood technology, no sliding components are used, so the load-bearing capacity, lubrication performance and the wear process are seldom interesting. On the other hand, the optical, coating and aesthetic properties are more important factors compared to the needs of the metal industry. Furthermore, some wood-working operations, such as planing, inherently produce a wavy surface which may also be an important roughness parameter.

In the 3D parameter set, the S_z profile height parameter is calculated from the five highest peaks and the 5 deepest pits. As a consequence, S_z is dominated by extreme values and it is hardly comparable with its R_z counterpart. At the same time, several other parameters such as S_a, S_q, S_{sk}, S_{ku}, S_{pk}, S_k and S_{vk} are more or less comparable with their 2D counterparts.

The weak correlation between S_z and S_q is given in Fig. 4.13 based on the results of two wood species. The S_z/S_q ratio varies between 10 and 18.

The same ratios for metal surfaces are between 6.5 and 11 (Dong et al. 1994).

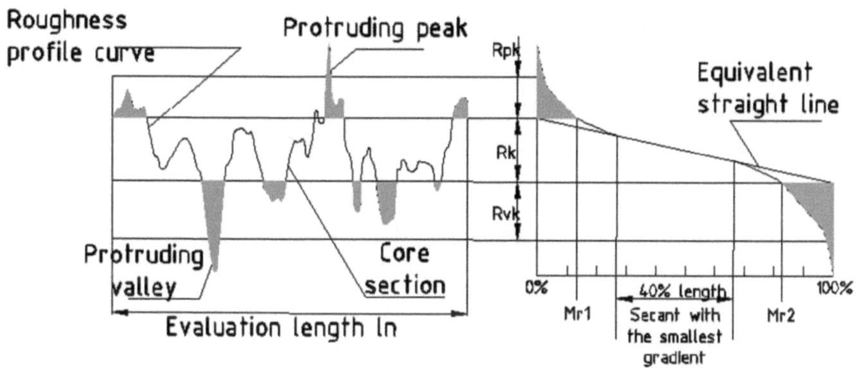

Fig. 4.12 The Abbott–Firestone curve and its related parameters

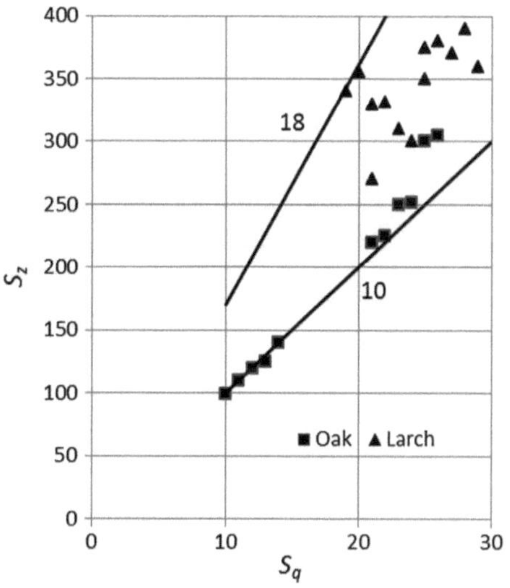

Fig. 4.13 Diagram of S_z versus S_q for two wood species

Some researchers in the metal industry have claimed that the arithmetic mean R_a (or S_a) and the RMS-based R_q are in correlation, and therefore, R_a can be expressed in terms of R_q. As a consequence, one of these parameters should be removed from the parameter set (Dong et al. 1994). Figure 4.14 shows the correlation between S_a and S_q for three wood species.

The S_q/S_a ratio varies in a wide range between 1.2 and 2.0. These results disagree with the above statement, and therefore, this question requires further study.

A peak protruding from the surface can be modelled in two dimensions by using the sine function raised to different powers, as shown in Fig. 4.15. Using high enough powers (5–15), differently spiked shapes can be achieved. The sine function

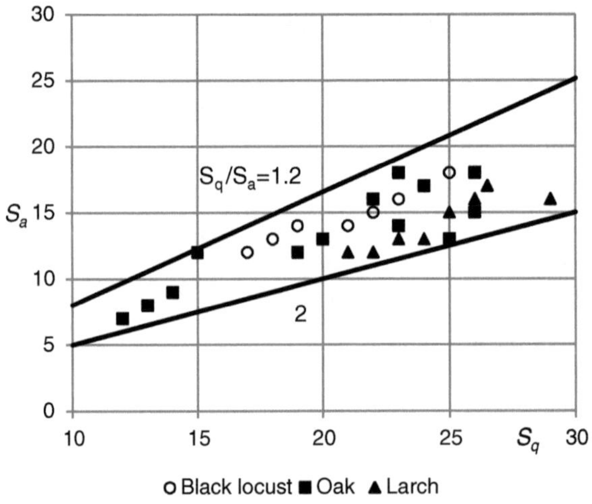

Fig. 4.14 Interrelation between the arithmetic mean S_a and root mean square S_q for different wood species

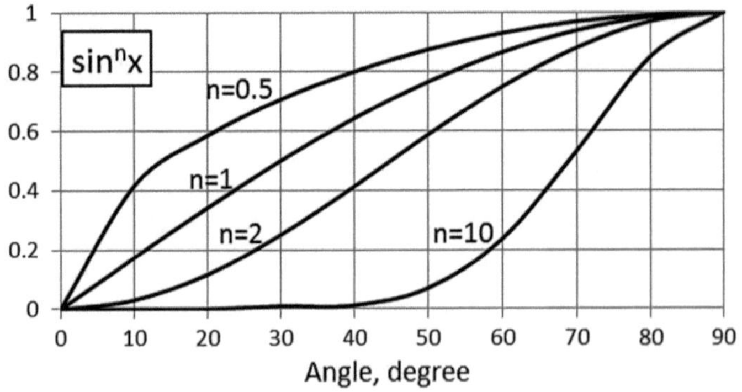

Fig. 4.15 Shapes of differently spiked sine wave

simplifies the calculation of the arithmetic mean R_a and its RMS value R_q. The arithmetic mean is simply integrated as

$$R_a = \frac{1}{\pi}\int_0^{\pi} \sin^n x \cdot dx \quad \text{if } n = 1, \quad \text{then } R_a = 2/\pi$$

which is related here to the unit height.

Performing the calculations for different exponents, a relation is obtained, which is represented in Fig. 4.16. The slenderness of the peaks considerably influences the R_q/R_a ratio which may be as high as 2.0 for real conditions in full agreement with the experimental results.

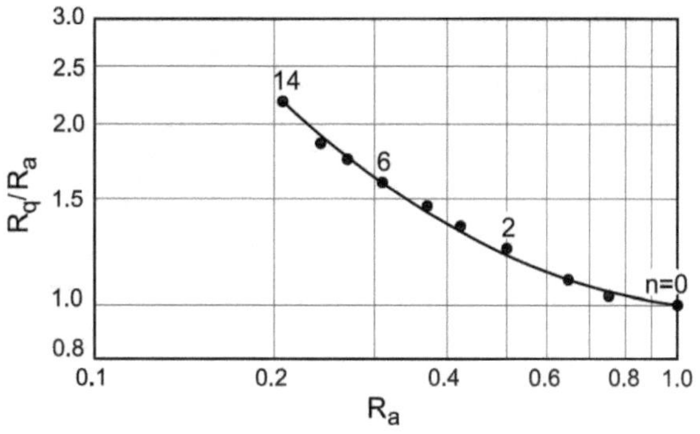

Fig. 4.16 Variation of the R_q/R_a ratio as a function of the average height of the differently spiked sine function

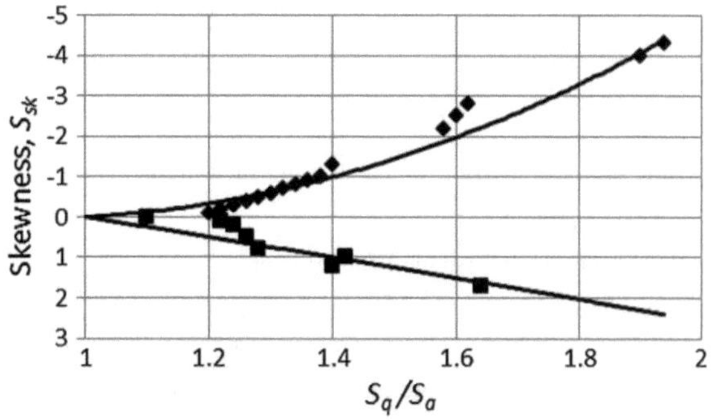

Fig. 4.17 Interrelation between skewness and S_q/S_a ratio for different wood species

Keeping in mind the above results, the R_q/R_a ratio (S_q/S_a ratio) may be used to characterise the slenderness of peaks and valleys. Namely, the correlation between R_q and R_a is not constant, as it is sometimes expected, but varies as a shape function of peaks and valleys.

Processing of 3D experimental results has shown that the S_q/S_a ratio has a strong correlation with the S_{sk} skewness. This relationship is given in Fig. 4.17. It is striking that the upper and lower spans are not fully symmetric. The relationship is valid for different wood species and differently manufactured surfaces. Most wood species having a more or less Gaussian height distribution are grouped around $S_q/S_a = 1.3$. Hardwood species (Oak, Black locust) have generally higher values.

In the metal industry, the relationship between skewness S_{sk} and kurtosis S_{ku} is well established (Dong et al. 1994). A similar function for wood species with differently manufactured surfaces is presented in Fig. 4.18. The correlation curve intersects the line of zero skewness at $S_{ku} = 3$ as it should for a Gaussian height distribution.

It is interesting to compare 3D and 2D measurement results on the same surface. Spruce and Oak samples were used in these experiments. A laser measuring unit was used for 3D measurements, and a conventional perthometer with 2 μm tip radius was used for 2D measurements. The results are given in Table 4.2.

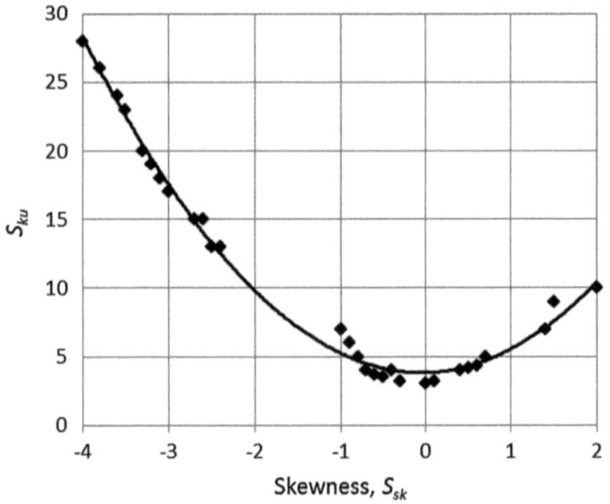

Fig. 4.18 Diagram of kurtosis versus skewness for different wood species and surfaces

Table 4.2 Comparison of 2D and 3D measurements

1. Spruce planed						
3D measurement, averaged for 10 spots						
S_{max}	S_z	S_a	S_{pk}	S_k	S_{vk}	$(S_{pk} + S_k + S_{vk})$
159.6	152.5	15.5	14.3	43	29.3	86.6
2D measurement, average of 30 line measurements						
R_{max}	R_z	R_a	R_{pk}	R_k	R_{vk}	$(R_{pk} + R_k + R_{vk})$
66.7	56.5	5.3	24.5	24.2	26.9	75.6
2. Oak planed						
3D measurement, averaged for 10 spots						
S_{max}	S_z	S_a	S_{pk}	S_k	S_{vk}	$(S_{pk} + S_k + S_{vk})$
390.6	373.8	20.1	26.5	45.2	83.1	154.8
2D measurement, average of 30 line measurements						
R_{max}	R_z	R_a	R_{pk}	R_k	R_{vk}	$(R_{pk} + R_k + R_{vk})$
107.8	71.5	6.4	36	26.9	43	105.9

According to expectations, S_z and R_z values are not comparable. S_a is roughly three times greater than R_a, while S_k is nearly 1.7 times more than R_k. At the same time, S_{pk} is only some 60 % compared to R_{pk}.

The sum of the three Abbott parameters is more or less comparable for Spruce. However, there is a discrepancy for Oak, probably due to the differences in valley heights. Further measurements are needed to clarify the real reasons causing these differences. The 3D measuring method rearranges the R_{pk} and R_k layers, and the laser beam can presumably measure the deep valleys more accurately.

Until now, the use of the height distribution parameters of skewness and kurtosis was not very common in the wood industry. Our opinion is that these parameters deserve more attention because they can provide useful information about the topography of wood surfaces. Here, we would like to refer to Sect. 4.10, where the use of these parameters made it possible to clarify the general behaviour of wetted surfaces.

4.5 Effect of Machining Operational Parameters on Roughness

Machining operations have fundamental effects on the surface topography independently of the type of materials. The main operational parameters influencing surface roughness of wood are the following:

- cutting speed,
- cutting depth,
- climb or counter cutting,
- oblique cutting,
- sharpness of the tool,
- feed per tooth,
- cutting circle radius,
- direction of cut,
- cutting angle,
- workpiece vibration.

The tool interacts with the timber material and its mechanical properties influence also the surface roughness. The main material properties are as follows:

- volume density,
- earlywood-and-latewood ratio,
- the hardness of wood,
- the modulus of elasticity,
- the anisotropy of wood,
- the homogeneity of wood structure.

It should be noted that some material properties are in loose correlation. It is generally true that a heavier timber has greater hardness and modulus of elasticity. Concerning surface roughness, the most important characteristic feature of timbers is their low tensile strength perpendicular to the grain. Excessive tensile forces causing presplitting may create a rough surface. Therefore, compressive forces should dominate in the close vicinity of the working edge.

Due to the free cutting mechanism (Dobler 1972), one of the most important influencing factors is the cutting speed. In order to achieve a smooth surface, the cutting speed should generally be between 30 and 50 m/s. A higher speed ensures not only a better surface but, at the same time, also increases the performance of woodworking operation per unit of time.

Figures 4.19 and 4.20 represent characteristic experimental results with Scots pine and Beech showing the roughness components (Magoss and Sitkei 2001). Examining the roughness components, by far the most varying component is the reduced valley height R_{vk}. Scots pine has no vessels, so the valleys in the height distribution originate from the cutting mechanism itself. At low cutting velocities, the inertia forces are not enough to counteract the edge force and the strength of the material is not high enough. Therefore, the counterforce and the force equilibrium can only develop at the expense of larger deformations in an extended volume causing deep outbreaks. The R_{vk} components steeply decrease up to cutting speeds of 50 m/s.

Essentially, a similar picture can be seen for Beechwood with some differences. Beechwood has a higher strength, and so it requires somewhat less inertia forces to ensure a clear cutting mechanism. Figure 4.20 clearly shows that around 30 m/s the counterforce is already well developed and a further increase in the cutting speed does not bring much improvement in the roughness values.

The improvements with cutting speed in the R_{pk} and R_k values are also more distinct for Scots pine compared to Beechwood.

The volume density of timbers largely determines their strength and hardness. As a consequence, heavier wood species have better surface roughness for the same woodworking operation. Figure 4.21 represents reduced peak height variation for

Fig. 4.19 Surface roughness parameters for Scots pine as a function of cutting speed

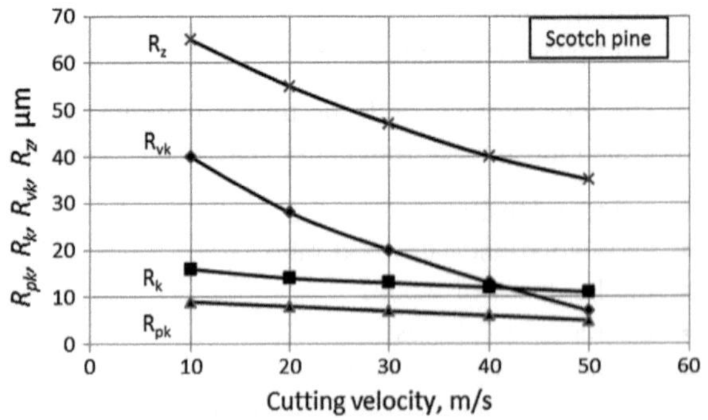

Fig. 4.20 Surface roughness parameters versus cutting speed for Beechwood

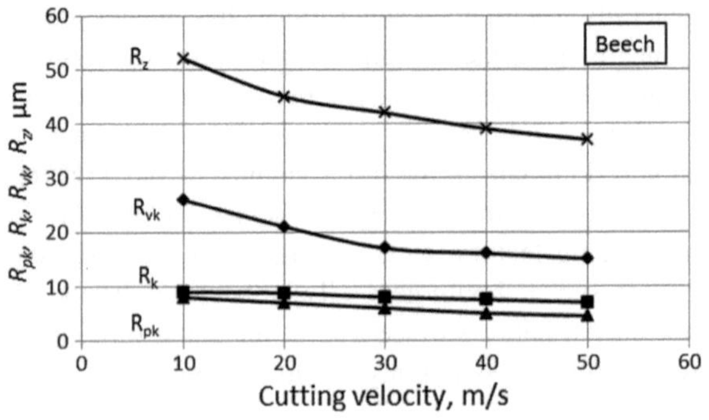

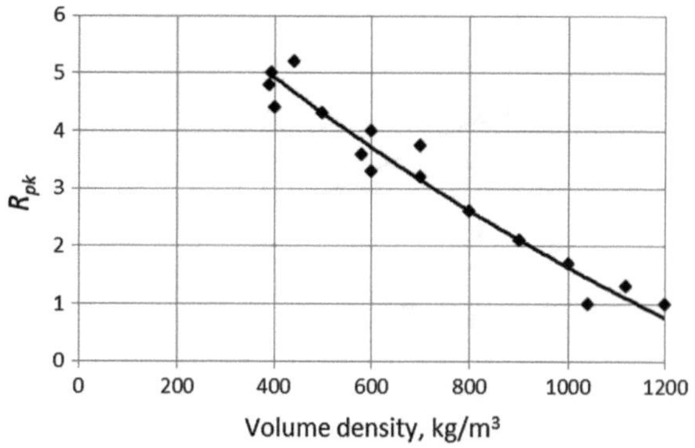

Fig. 4.21 Reduced peak height as a function of volume density for different wood species

different wood species as a function of volume density. The experimental results show some scattering due to species variations, but the correlation is definite. Over 1000 kg/m^3, the reduced peak height R_{pk} decreases to values as low as 1 µm.

The measurement of roughness parameters is generally carried out perpendicular to the grain. It is interesting to know the variation of roughness measured with different trace angles to the grain. These measurement results are given in Fig. 4.22 for Scots pine and Oak (Magoss and Sitkei 2001). The lowest roughness is given for tracing along the grain. It is interesting that maximum values are obtained at intermediate tracing angles around 60°. The ratio of the value at 90° and the minimum values at zero degrees may be quite different for various wood species. Oak has a ratio several times larger than Scots pine (see Fig. 4.22).

Other roughness parameters may have somewhat different behaviour as a function of trace angle to the grain. Figure 4.23 shows measurement results of R_{pk}, R_a and R_k/R_z ratio for Scots pine. R_{pk} and R_a roughness parameters have an explicit maximum at trace angles around 50°. At the same time, the R_k/R_z ratio has somewhat higher values at small tracing angles.

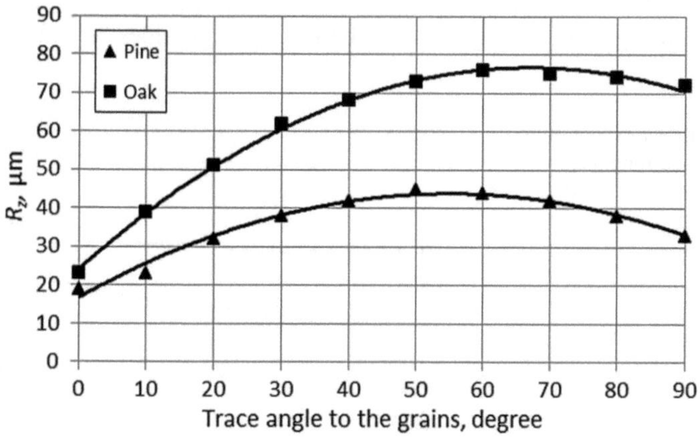

Fig. 4.22 Roughness of Scots pine and Oak as a function of trace angle to the grains

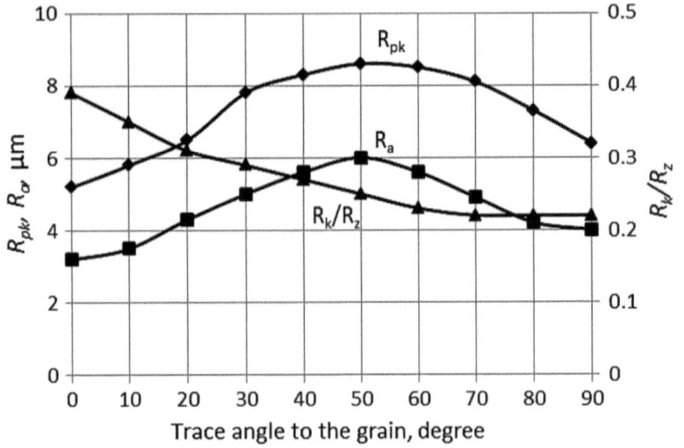

Fig. 4.23 Change of roughness parameters as a function of trace angle to the grains. Scots pine

The feed per tooth has also some effect on the surface roughness. It is well known in practice that the very essence of a smooth machining is the small feed per tooth as the last operating sequence. A small tooth bite means lesser cutting forces and lesser deformation and damage in the separation zone.

Figure 4.24 represents roughness variations for three wood species as a function of tooth bite (Magoss 2008). The curves can be described with the following empirical equation:

$$R_z = A + B \cdot e_z^n$$

where e_z means the feed per tooth in mm, the exponent n for all three wood species may be taken as 0.6, and the constant B has a value of 43 (R_z is given in µm) for all wood species. The meaning of constant A is interesting. It is surprisingly very close to the minimum roughness of wood species corresponding to the anatomical roughness in Fig. 4.33. In Fig. 4.24, the best fit for A provides the values of 69, 35

Fig. 4.24 Effect of tooth bite on the average roughness height R_z for three wood species

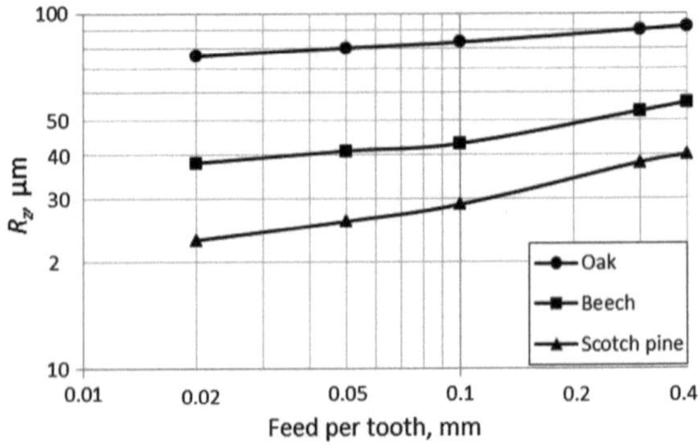

and 19 for Oak, Beech and Scots pine, respectively. Their anatomical roughness values shown in Fig. 4.33 are 65–75, 35–40 and 20 which are in close agreement with the experimentally obtained values. This finding allows an approximate estimation of roughness for other wood species.

An interesting but very seldom investigated question is workpiece vibration and its effects on surface roughness. A vibrating system that consists of mass and spring always vibrates if excitation takes place. During woodworking operations, the radial component of the cutting forces performs excitations on the workpiece. Two different cases may be distinguished in different woodworking machines: the workpiece is supported by a rigid surface (e.g. machine table) or it is supported by press rolls from above against the exciting forces (Sitkei et al. 1990). The spring constant in the first case is proportional to the modulus of elasticity of the timber. In the second case, it equals the spring constant of press rolls. New experiments have shown that the pneumatic clamping of a workpiece cannot be regarded as a rigid one (Csanády and Magoss 2013). In certain cases, the vibration amplitudes may not only influence the surface roughness but also cause unequal tooth feed and cutting force using multi-edge tools.

Experiments on a planer and thicknesser have shown that the periodic transverse displacement of the workpiece fundamentally depends on its support. The direction of exciting forces determines the workpiece deflection, whether they are directed towards the machine table or the supporting rolls. Figure 4.25 shows experimental results carried out on a planer and thickener with Scots pine samples (Sitkei et al. 1990). Due to the continuous wear of the tool, the radial component of the cutting force will increase as a function of working hours. As a consequence, the rising exciting force further increases the workpiece amplitudes.

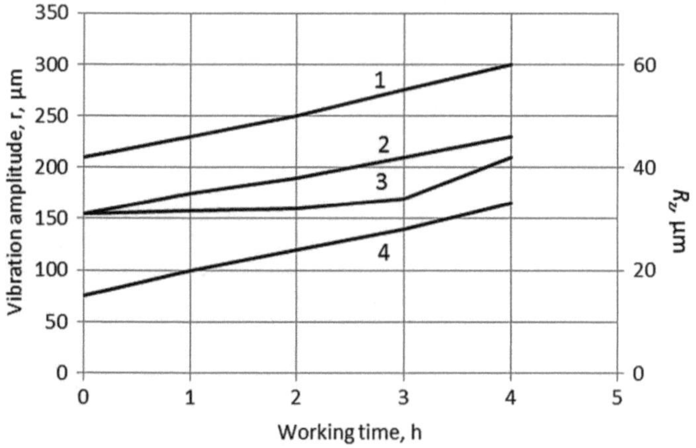

Fig. 4.25 The effect of working hours on the vibration amplitudes and surface roughness R_z. Scots pine. 1-R_z, planer; 2-*r*, planer; 3-R_z, thicker; 4-*r*, thicker

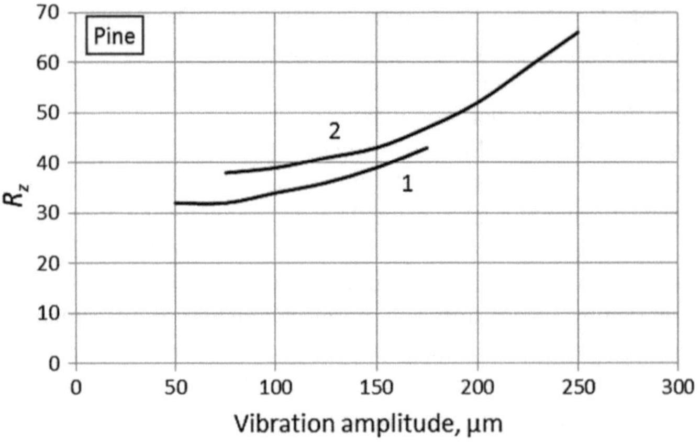

Fig. 4.26 Influence of vibration amplitudes of the workpiece on the surface roughness. *1*—sharp tool, *2*—blunt tool

The periodic deflection of the workpiece interacts with the tool edges and produces increased roughness. The deflection and roughness will considerably be higher for a planer due to the lower spring constant of the supporting rolls.

The appropriate processing of experimental results enabled us to find a direct relationship between vibration amplitudes and the surface roughness parameter R_z, which is depicted in Fig. 4.26. It means that vibration of the workpiece should be controlled as much as possible.

A sign of tool vibrations is washboarding mainly occurring with bandsaws and circular saws, but it was also observed with frame saws in certain cases (Okay et al. 1995; Lehmann and Hutton 1997; Orlowski and Wasielewski 2001). Washboarding is a wavelike surface pattern appearing on the surface of sawn timber, as shown in Fig. 4.27. The washboarding profile can be as high as 0.6–0.8 mm deep on the face of the lumber requiring a larger target size to remove sawing deviations and the

Fig. 4.27 Typical wavelike pattern of a washboarding on bandsaws

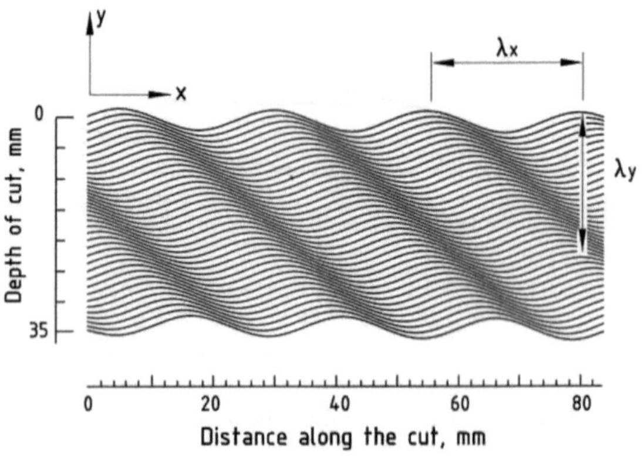

depth of washboard. Washboarding primarily appears when higher blade speed, feed speed and thinner bandsaw and circular saw blades are used.

Washboarding is a self-excited resonance phenomenon with an important condition: the tooth passage frequency should slightly be higher than the natural frequency of the blade (Okay et al. 1995). It is also known that all washboarding with bandsaws was produced by torsional vibration modes.

Washboarding occurs in a narrow frequency interval, approx 4 % of the natural frequency. It can generally be eliminated by increasing or decreasing the blade speed. A similar effect results from increasing or decreasing the bandsaw tension which alters the natural frequency of the saw blade. Furthermore, lowering the tooth bite results in smaller excitation forces reducing the severity of washboarding.

Special attention should be paid to edge machining of boards which are very sensitive to edge breaking. Edge breaking can be reduced by using a larger diameter tool and very thin chips. This may be accomplished by using cone-shaped milling cutters with a setting angle as depicted in Fig. 4.28 (Кряжев 1963; Lang 1989; Tröger and Lang 1990). In this case, the knife-edge describes a hyperbolic path and its virtual cutting circle radius may have enormous values depending on the setting angle.

The cutting radius is given by the following equation:

$$R_v = L \cdot \tan \alpha$$

and

$$L = \frac{D_a}{2 \cdot \cos \chi} = \frac{D_a}{2 \cdot \sin \alpha} \qquad \alpha + \chi = 90^\circ$$

where

D_a is the mean diameter of the tool (see Fig. 4.28),

χ is the setting angle, generally around 5°.

Taking $\alpha = 85°$, $L \cong D_a$, $\tan 85° = 11.43$ and that means that the virtual radius of cutting is eleven times larger than the mean radius of the tool.

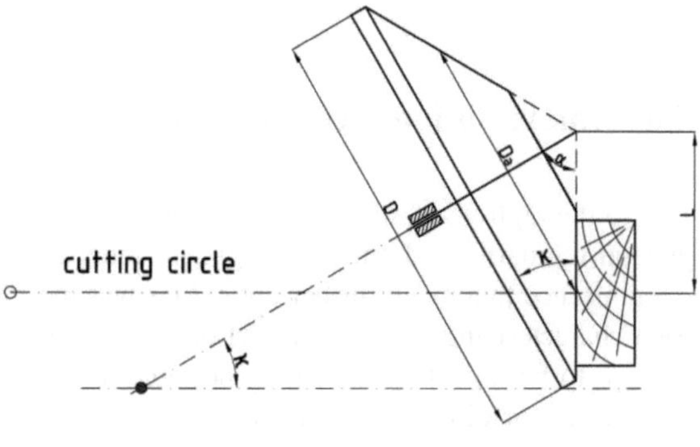

Fig. 4.28 Working principle of a cone-shaped milling cutter

The chip thickness is calculated as

$$h = e_z \cdot \sin\varphi \cdot \sin\chi$$

where
e_z is the feed per tooth,
φ is the tool rotation angle under cutting.

Due to the small value of $\sin\chi$ (0.07–0.1), the chip thickness may be very thin, even by using higher feed per tooth.

The wave depth (geometric roughness) t_1 is obtained from the equation

$$t_1 = \frac{e_z^2}{8 \cdot R_v}$$

which may also be very small due to the large cutting radius. Experimental results supported the above statements, and using setting angles in the range of 5°–8°, the tool can work with tooth bites up to 5 mm and practically without edge breaking (Lang 1989; Tröger and Lang 1990).

Finally, we briefly consider the remaining influencing factors. The climb cutting with a sharp edge generally produces a somewhat better surface quality compared to counter cutting. The reason is that the edge always exerts compressive stress onto the material. Another explanation may be the continuously decreasing chip thickness towards the exit of the edge of the material.

The cutting angle (or rake angle) also influences surface roughness. Cutting angles around 65°–70° exert more compressive stresses onto the material which produces a smooth surface. It is well known that the use of a small cutting angle in cutting veneer requires a pressure beam to provide the necessary compressive stresses.

Oblique cutting, or slide cutting, utilises the combined effect of compressive and shear stresses to induce rupture. This cutting requires less compressive forces and produces less deformation. For instance, very thin veneers can only be cut by using large oblique angles (50°–60°). The reduction of compressive forces and associated deformations generally provides slightly less surface roughness (Sitkei 1997).

4.6 Interrelations Among Roughness Parameters

Due to the high variability of wood structures, it took a long time to discover some general regularities of surface roughness independently of wood species. Another circumstance that made further difficulties was the myriad of roughness parameters, each characterising only one aspect of a given surface. Under the circumstances, it seemed to be useful to search for possible relationships among surface roughness parameters for different wood species.

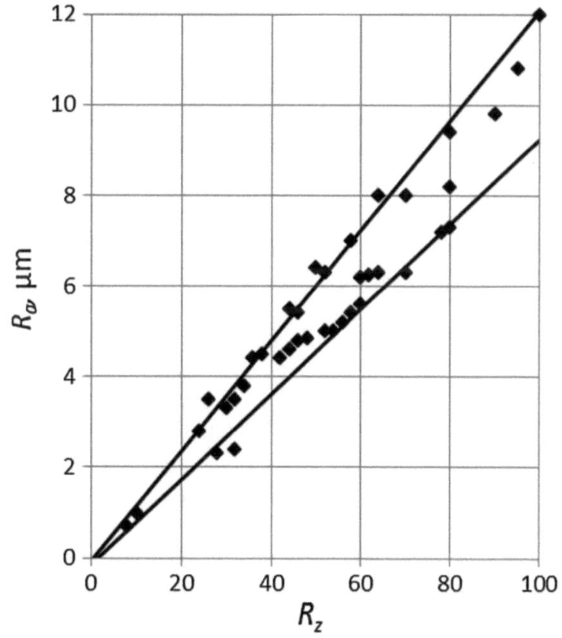

Fig. 4.29 Diagram of R_a versus R_z for different wood species

In the search for new ways to generalise the existing experimental results and to recognise and establish general laws, the elaboration of a structure number uniquely characterising each different wood species in relation to its roughness is proved to be very useful. It is treated in details in the next section.

The average roughness R_a is linearly correlated with the irregularity depth R_z as depicted in Fig. 4.29. Their ratio varies between 0.09 and 0.12.

The bigger ratios are generally related to wood species with wider voids (vessels). R_a is dependent on the size of voids, while R_z depends only on the height of the irregularities.

It is interesting that R_a has a better correlation with the sum of the Abbott parameters (R_{pk} + R_k + R_{vk}) (Magoss and Sitkei 2001). The correlation obeys the following empirical equation:

$$R_a = 0.275 \cdot \left(R_{pk} + R_k + R_{vk}\right)^{0.85}$$

The roughness height R_z correlates with the sum of Abbott parameters in a similar manner:

$$R_z = 2.41 \cdot \left(R_{pk} + R_k + R_{vk}\right)^{0.85}$$

and from the above equations, it follows that

$$R_a = 0.114\, R_z$$

which is the averaged straight line in Fig. 4.29.

A further examination of the scattering zone in Fig. 4.29 has revealed that the reason for this is the possible variation of R_k which is the major component of R_a. To discover the essence of scattering, Fig. 4.30 represents the variation of the R_a/R_z ratio as a function of the R_k/R_z ratio for different wood species. An increase of the R_k/R_z ratio within a given species always enlarges the R_a/R_z ratio. For a given R_k/R_z ratio, hardwoods with big vessels always have a higher R_a/R_z ratio compared to conifers.

The curves in Fig. 4.30 can be described with the following empirical equation:

$$\frac{R_a}{R_z} = \text{const} \cdot \sqrt{\frac{R_k}{R_z}} \quad \text{or} \quad \frac{R_a}{\sqrt{R_k \cdot R_z}} = \text{const.}$$

and this dimensionless number may be regarded as a similarity number of the surface roughness.

The above relations can also be interpreted for three-dimensional measurements. Due to the definition of S_z, it does not correspond to R_z, and therefore, a similar relation for S_z cannot be established. At the same time, S_a and the Abbott parameters S_{pk}, S_k and S_{vk} are fully useful. Their correlation is shown in Fig. 4.31, and the averaged correlation is given by the equation:

$$S_a = 0.268 \cdot \left(S_{pk} + S_k + S_{vk}\right)^{0.85}$$

which practically agrees with the corresponding R_a function.

Using the Abbott parameters, the lack of material in the surface profile can be determined. An equivalent layer thickness corresponding to the voids can be defined in the following manner:

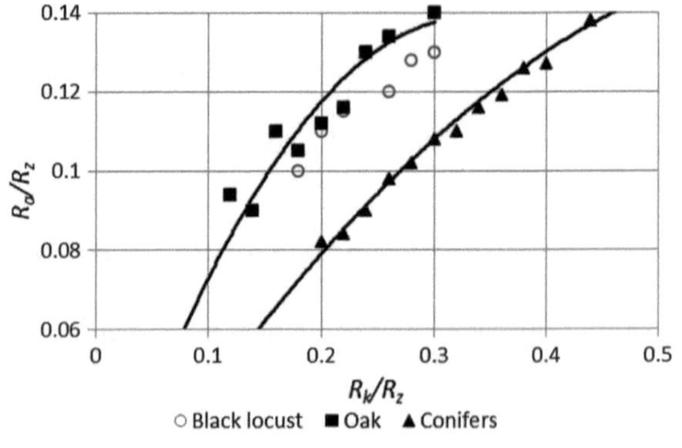

Fig. 4.30 Variation of the R_a/R_z ratio as a function of R_k/R_z ratio for different wood species

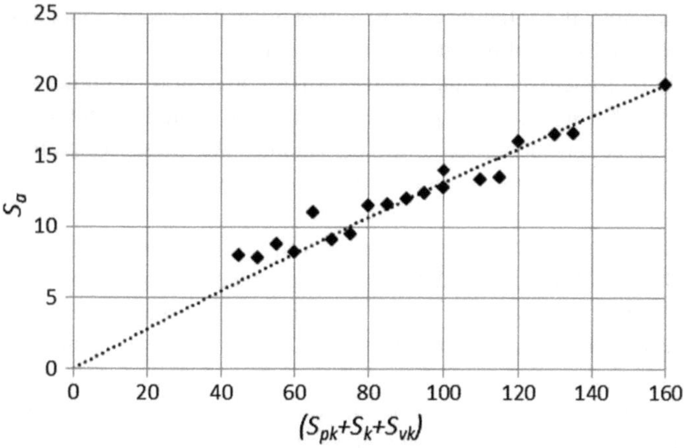

Fig. 4.31 Variation of S_a as a function of Abbott parameters for different wood species

$$h = R_{pk} \cdot \left(1 - \frac{M_{r1}}{2}\right) + \frac{R_k}{2} + R_{vk} \cdot \left(1 - \frac{M_{r2}}{2}\right)$$

For the three-dimensional case, the same equation is valid, replacing the two-dimensional parameters with the corresponding three-dimensional ones.

Figure 4.32 shows the lack of material in the surface profile related to the unit surface area, using three-dimensional measurement results. It is interesting and important to note that the earlier measurements with two-dimensional perthometer gave significantly smaller values, although the courses of the curves were quite similar to those given in Fig. 4.32 (Magoss and Sitkei 2001). The determination of the lack of material is a three-dimensional problem, and in this respect, the three-dimensional measurement should give more accurate results. In many other cases, however, the two- and three-dimensional measurements give comparable

Fig. 4.32 Lack of material in the surface profile as a function of Abbott parameters for different wood species. *1*—including R_{pk}, *2*—without R_{pk}

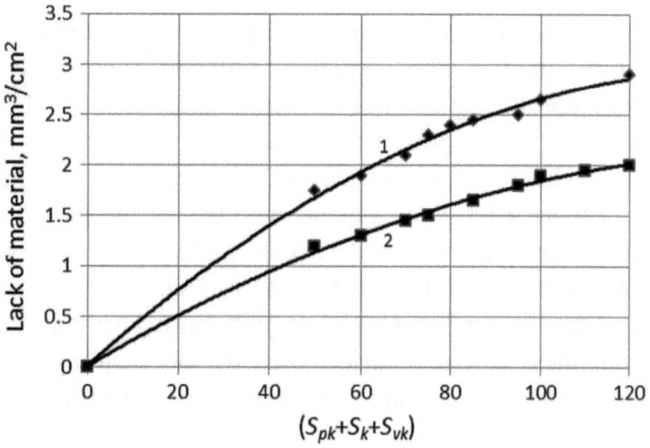

results. Naturally, the reliability of two-dimensional measurements can considerably be enhanced by multiple tracing side by side.

It should be remembered that several further correlation functions such as $S_a = f(S_q)$, $S_{sk} = f(S_q/S_a)$ and $S_{ku} = f(S_{sk})$ have already been treated in Sect. 4.4. Additional relations will be given in Sect. 4.8 for sanding. All these correlation functions highly facilitate the interpretation of measurement results. It is possible to transform two-dimensional measurement results into three-dimensional and vice versa. Knowing different correlation functions, even without measurements, an appropriate prediction of an expected roughness parameter becomes possible.

In the past, researchers and application engineers accustomed the two-dimensional roughness parameters and a given value has been felt clear in its valuation.

For the future, it is very important to elaborate a correspondence between two-dimensional and three-dimensional results and to give a clear explanation for the existing differences.

4.7 The Use of Structure Number

The structure number ΔF, introduced in Sect. 4.2, gives a unique characterisation of a particular wood species in relation to its internal structure which determines the inherent roughness due to its anatomy. The need for a structure number arose from the fact that the term "wood species" as a variable cannot be treated numerically, and therefore, different wood species cannot be compared. Although some general relationships may exist among various wood species, it is impossible to recognise them without treating wood species as a variable.

Large differences may exist in the internal structure of a given wood species. For instance, Scots pine growing in northern Europe has a higher density than the same species growing in Central Europe. The structure number can take into account such differences within species. The structure number is thought to have a definite correlation with roughness parameters, regardless of the wood species and the location where they were grown.

Knowledge of the structural properties of different wood species is required in order to use the structure number. The determination of these data is time-consuming and this is a drawback. Knowing these structural properties, it was possible to recognise some general regularities of surface roughness. Furthermore, it provides guiding principles for comparing roughness of different wood species.

Figure 4.33 gives a clear overview of the possible values of surface roughness for quite different wood species milled with cutting velocities of 10 and 50 m/s (Magoss and Sitkei 2003). It also shows the theoretically possible minimum roughness component. Using high cutting velocities and sharp tools, the resultant roughness is mainly determined by the anatomy of woods and the woodworking operation adds a smaller component of the resultant value.

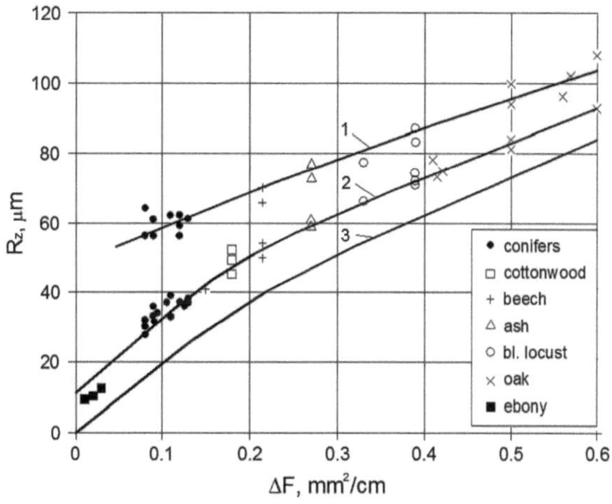

Fig. 4.33 Average roughness R_z as a function of structure number ΔF. *1*—cutting speed 10 m/s; *2*—cutting speed 50 m/s; *3*—anatomical roughness

This is especially true for hardwood species with large vessels. Conifers have less anatomical roughness, and their roughness due to machining may be the same magnitude as that caused by their anatomy.

Conifers with less own strength are more sensitive to lower cutting speeds compared to hardwoods.

Using the structure number, it was possible to represent all wood species in a diagram with continuous curves and to recognise the general behaviour of different wood species according to their internal structure. There may be smaller deviations due to particular properties not treated here.

The results presented in Fig. 4.33 are valid for planed and milled surfaces using sharp knives, but definitely not valid for sanded surfaces. Sanding will be discussed in detail in the next section.

The curves given in Fig. 4.33 can be expressed with the following empirical equation:

$$R_z = \left(123 \cdot \Delta F^{0.75} + 35 \cdot e_z^{0.6}\right) \cdot \left(1 + \frac{50 - v_x}{50} \cdot \frac{0.1183}{\Delta F^{0.83}}\right)$$

$$10\,\mathrm{m/s} \leq v_x \leq 50\,\mathrm{m/s}$$

where ΔF must be substituted in mm^2/cm, e_z in mm and v_x in m/s.

The third part of the above equation takes the smaller strength of softwoods into account.

Several roughness parameter ratios can be constituted uniquely depending on the structure number ΔF. The results illustrated in the following figures refer to milled and sanded surfaces. The results due to sanding are mainly valid for sandpaper grades between P-80 and P-150.

Figure 4.34 represents the R_a/R_k ratio for milled and sanded surfaces as a function of their structure number. This ratio shows the relative contribution of the

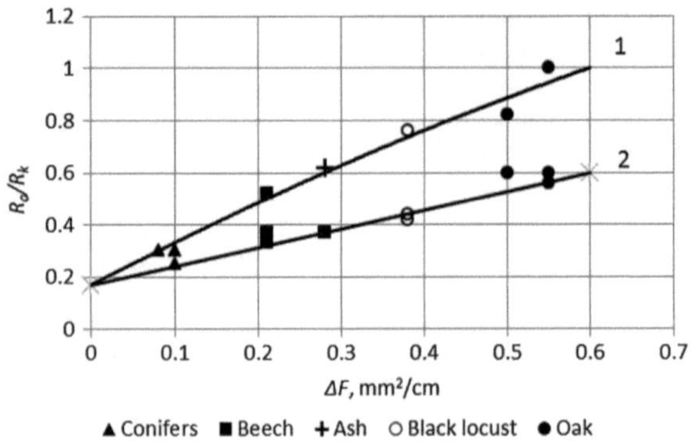

Fig. 4.34 Variation of R_a/R_k ratio as a function of structure number. *1*—milled, *2*—sanded

R_k layer to the average roughness. Conifers lacking deep valleys have a stronger dependence on the R_k layer thickness. Sanding a surface restructures the height distribution and increases the role of the R_k layer, especially with hardwood species.

Figure 4.35 shows the variation of R_k/R_z ratios as a function of structure number. This ratio for planed or milled conifers is roughly 1/3, which decreases to 0.1 for hardwood species with large vessels.

The restructuring effect of sanding can more clearly be seen here increasing the R_k/R_z ratio considerably for all wood species without exception. Maximum deviation occurs also here for big vessel hardwood species.

The relative valley height R_{vk}/R_z is depicted in Fig. 4.36 for milled and sanded surfaces. This ratio for conifers is about 0.25 and it increases up to 0.7 for big vessel species. The sanding lowers this ratio due to compaction and clogging effect. The rate of lowering is slightly increasing toward big vessel hardwood species.

In the previous section, a roughness similarity number in dimensionless form was introduced which is represented in Fig. 4.37 as a function of structure number ΔF. The figure summarises the measurement results for different wood species using sharp and blunt tools, different cutting speeds, planing, milling and sanding. The experimental results show a given scattering and a slight average increase as a function of structure number. A good estimation for planed or milled surfaces made by sharp tools is given by the relation:

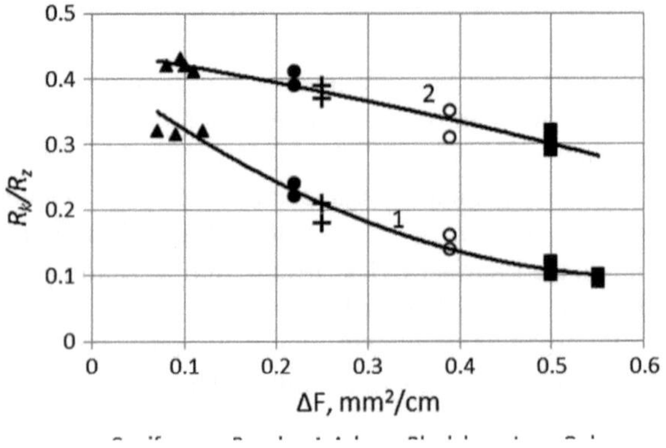

Fig. 4.35 R_k/R_z ratio versus structure number. *1*—milled, *2*—sanded

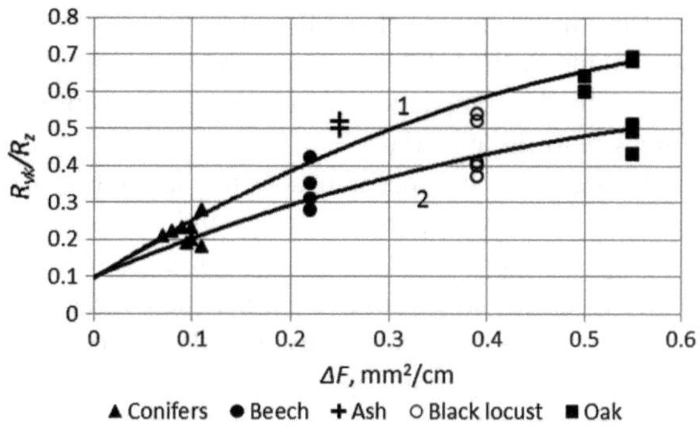

Fig. 4.36 Variation of R_{vk}/R_z ratio versus the structure number. *1*—milled, *2*—sanded

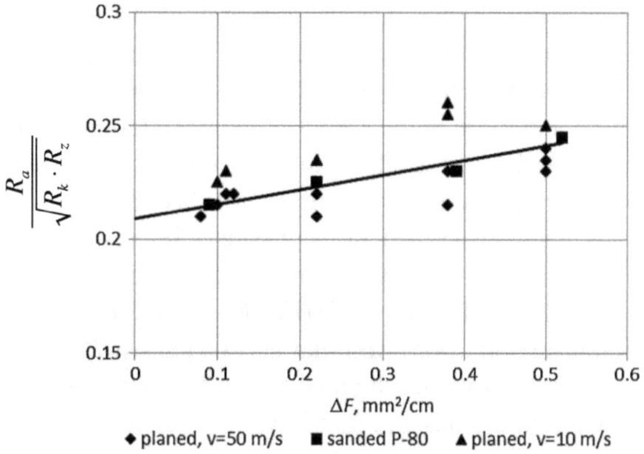

Fig. 4.37 Similarity relationships for variously manufactured wood surfaces

$$\frac{R_a}{\sqrt{R_k \cdot R_z}} = 0.225$$

for all wood species.

Using low cutting speeds and sanded surfaces, for the constant somewhat higher values are to be expected, especially for species with large vessels (see Fig. 4.37).

The use of similarity numbers highly facilitates the generalisation of a vast amount of experimental results for different wood species and woodworking operations.

4.8 Surface Roughness at Sanding

Sanding is a very common woodworking operation to produce smooth surfaces or to adjust a given thickness (e.g. particle board). It is widely accepted that surface sanding is proved to be the most advantageous processing step prior to coating and painting (Richter et al. 1995 and see Sect. 4.10).

Due to its almost spherical cutting edge, sanding is an unusual cutting process with some distinct differences compared to knife cutting. The main differences are caused by the compaction and clogging effects of grits always exerting compressive forces onto the material.

In Sect. 4.3, the mechanics of the sanding process was treated in detail and a comparison between knife cuttings and sanding was also presented. An important finding was that a knife with a given edge radius compacts the underlying layer twice as deeply as grit does with the same radius. Another important conclusion is that the compaction exerted by the edges is directly related to the core depth R_k.

The most important operational parameters of sanding are the cutting speed, surface pressure, grit size and feed speed. They determine the specific wood removal (g/m^2·min) and the surface roughness parameters for a given wood species. It is well known that the specific wood removal continuously decreases as a function of working time due to the wear process. The linearity between stock removal and surface pressure in a wider range does not hold. At higher pressures, more and more chips are in the contact zone hindering the penetration of grits into the material and resulting in a decrease of wood removal. Similarly, wood removal is theoretically independent of grit size. However, using fine grits, wood removal may decrease considerably. Readers interested in the relationships describing the interaction of the operational parameters are referred to the special literature (Csanády and Magoss 2013; Magoss 2013; Dobrint et al. 1991; Scholz and Ratnasingam 2005; Siklinka and Ockajova 2001).

In the following, the results of a large-scale experimental work will be presented to explain the basic regularities of the sanding process. Five different wood species (Spruce, Larch, Beech, Black locust and Oak) were selected with known structural properties (Magoss and Sitkei 2000). In these experiments, four grit sizes (P-80, P-120, P-150 and P-240) were used with sanding velocities around 25 m/s and with surface pressure between 0.25 and 0.35 N/m^2. Ten measurements were taken on each sample using a perthometer with a tip radius of 2 μm.

The average roughness R_a as a function of the sum ($R_{pk} + R_k + R_{vk}$) is depicted in Fig. 4.38 using all results of the wood species tested. The curves show a picture

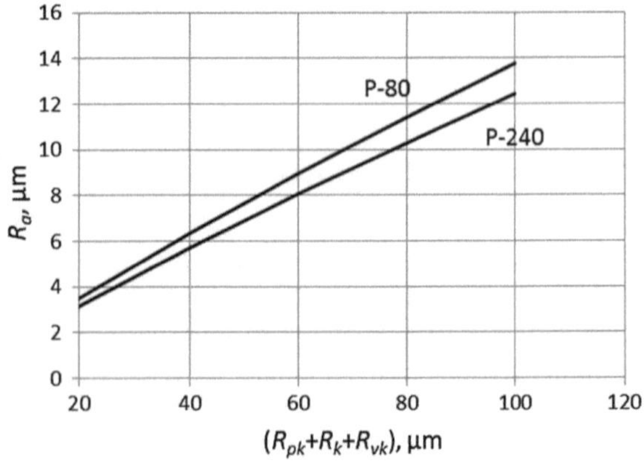

Fig. 4.38 The average roughness R_a as function of the sum ($R_{pk} + R_k + R_{vk}$)

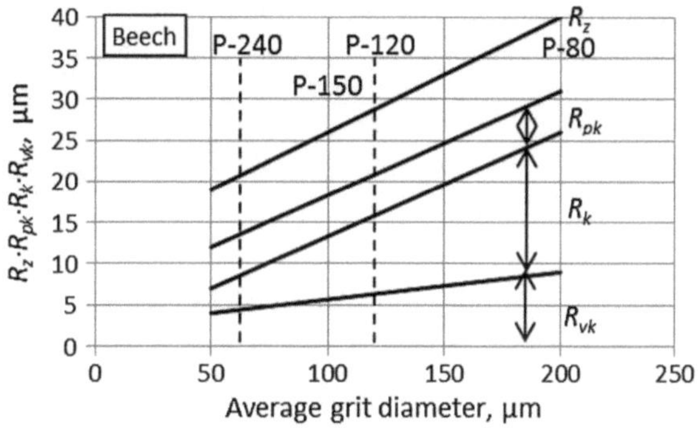

Fig. 4.39 Effects of grit size on the surface roughness parameters for Beechwood

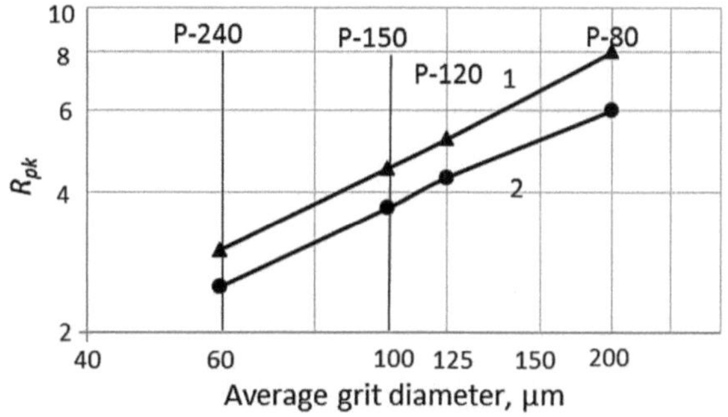

Fig. 4.40 Relationship between the reduced peak heights R_{pk} and grit size

quite similar to that of machined surfaces (Magoss and Sitkei 2001). Fine sanding with P-240 grit size results in somewhat lower average roughness. Concerning R_a, sanding with P-80 approximately corresponds to a surface machined with knife. The curves given in Fig. 4.38 can be described in a similar way as those in Sect. 4.5 for a knife-machined surface.

A detailed roughness representation for Beechwood is given in Fig. 4.39 as a function of average grit diameter. The relationship is almost linear for all roughness components. The strong dependence of the R_k layer is striking. This finding refers to the strong influence of the particle diameter on the crushing depth in the surface layer.

The smoothness of a surface is considerably determined by the reduced peak heights. This relationship for Spruce and Beech is given in Fig. 4.40 on a double logarithmic scale. The decrease of grit size creates quite low values which would give a good polishing possibility using wax or light-coloured oils. The polishing ability of a surface plays a definite role in the colour enhance of a given wood species (Sitkei 2013).

The core depth R_k should have a strong correlation with the grit size. Figure 4.41 shows the R_k/R_z ratio as a function of structure number ΔF. The relationship is quite similar to that of machined surfaces, and the P-240 curve almost coincides with the curve valid for machined surfaces. Here, we can make an interesting comparison: the average edge radius for sharp knives is 20 μm, while the average grit radius for

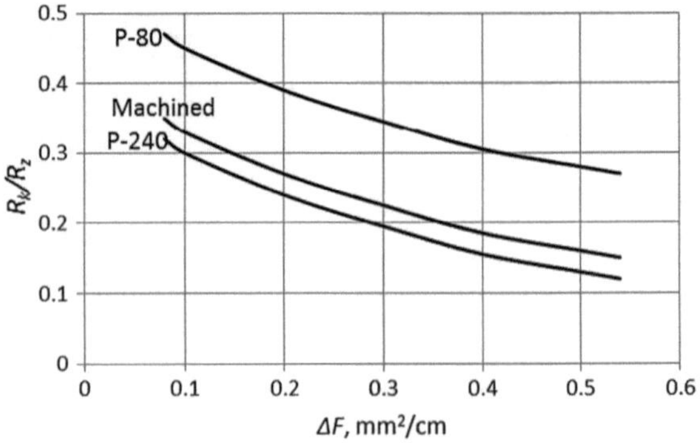

Fig. 4.41 Relative core depth as a function of structure number ΔF

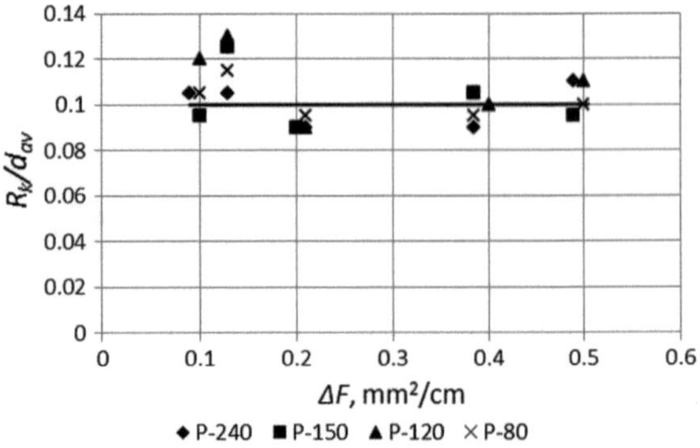

Fig. 4.42 The ratio of core depths and grit diameter depending on the structure number ΔF

P-240 is 30 μm. This experimental result is in full agreement with Fig. 4.9 in Sect. 4.3 due to the different action of a sphere and a cylinder on the half-space.

Another interesting finding is that the core depth R_k related to the average grit diameter is constant for all wood species. This relationship is shown in Fig. 4.42, which makes it possible to forecast the expected R_k layer in advance or to select an appropriate grit size to achieve a given R_k layer thickness.

The reduced valley depth R_{vk} related to the R_z value shows a strong correlation with the average grit diameter. Figure 4.43 demonstrates this relationship for the wood species tested. Spruce, Larch and Beech have similar values, while Black locust and Oak show significantly higher R_{vk} values.

An overall view on the irregularity depth R_z for all wood species is given in Fig. 4.44, showing also the minimum anatomical roughness.

Measurement results have clearly shown that the average roughness R_z may be considerably less than the anatomical roughness (Magoss and Sitkei 2000, 2001) for a given wood species. This is possible only if the surface layer undergoes a structural change due to cell deformation. Besides the crushing effect of spherical edges, a clogging effect (compression) takes place, decreasing the cavities in number and size which clearly appears in the roughness measurement results.

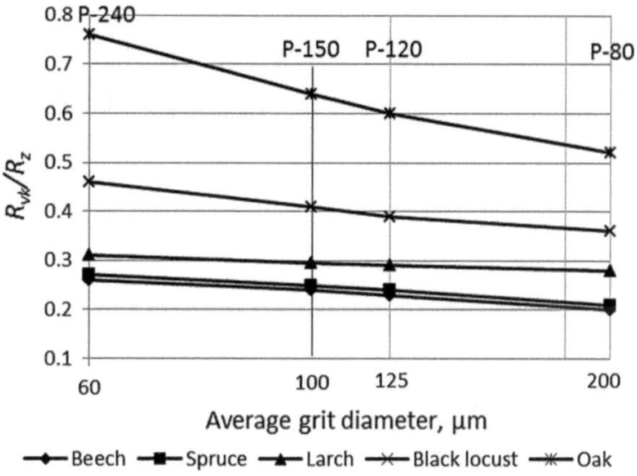

Fig. 4.43 The R_{vk}/R_z ratio as a function of grit size

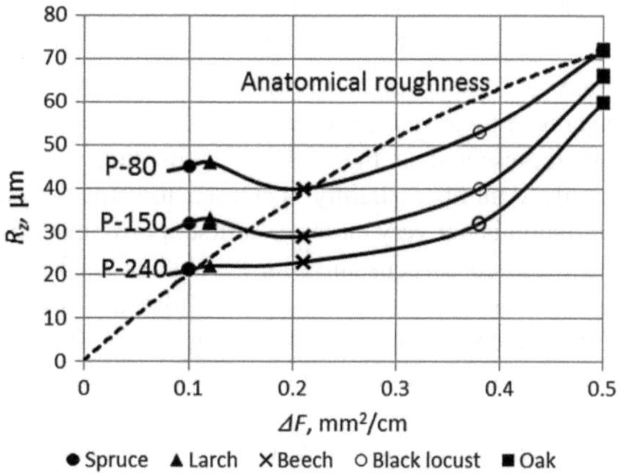

Fig. 4.44 Overview of R_z values for all wood species as a function of structure number ΔF

The local compression of material under a working edge depends on the local strength and not on the overall strength of the material. The local strength of wood materials is not known today, but estimation for comparison is possible using the structural properties of the wood materials in question. These structural properties are given in Table 4.1 in Sect. 4.2.

Beech seems to be very sensitive to the compression effect of the edge. Looking at its structural properties, we can see the very large number of tracheids both in the earlywood and in the latewood. A large number of cavities in a given cross section mean thinner cell walls which can be more easily deformed and compressed, decreasing the bigger cavities. As a consequence, a clogging effect due to surface layer deformation occurs.

Comparing Black locust and Oak, we can see a similar case: there are two to three times as many tracheids in Black locust in the earlywood and latewood compared to those of the Oak. Therefore, Oak will be clogged much less than Black locust, in spite of the fact that the structure numbers of Black locust and Oak are not

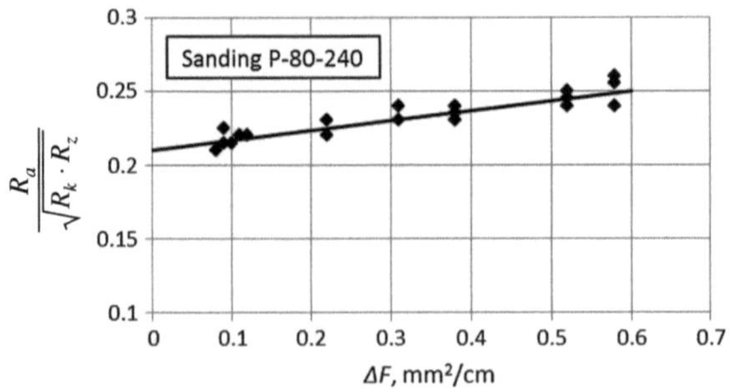

Fig. 4.45 Dimensionless plot of experimental results of sanding using different sanding grades and wood species

far apart. As a consequence, Black locust shows a much smoother surface compared to Oak after the same sanding.

In the previous section, Fig. 4.37 shows that the dimensionless quantity $\frac{R_a}{\sqrt{R_k \cdot R_z}}$ is more or less invariant for different wood species and machining operations. The averaged measurement data for sanding using grit grades between P-80 and P-240 are represented in Fig. 4.45 as a function of structure number ΔF. The dimensionless number slightly increases towards species with large vessels, but the variation is not very large. As a first approximation, the dimensionless number may be regarded as constant with a value of 0.23. This finding highly facilitates the generalisation of experimental results and evaluation of the sanding process. Furthermore, a comparison of Figs. 4.37 (Sect. 4.7) and 4.45 reveals that the derived dimensionless number is invariant independently of machining operations and wood species.

Based on the theoretical considerations and experimental results, the following main conclusions may be drawn

- between the standard grit size P and the average grit diameter a hyperbolic relationship exists,
- the stress under a grit, as a spherical edge depends on the surface pressure and the modulus of elasticity of the wood material, but does not depend on the grit size,
- it is proved theoretically and experimentally that the core depth R_k strongly depends on the average grit diameter used,
- experimental results show that all three components of the Abbott curve (R_{pk}, R_k, R_{vk}) are nearly linear, increasing with greater grit diameters,
- an interesting finding that the R_k/d_{av} ratio remains constant for all wood species tested,
- related roughness parameters (R_k/R_z, R_{vk}/R_z) show definite relationships as a function of either the structure number ΔF or the average grit diameter,

- experimental results show that the roughness after sanding may be smaller than the anatomical roughness of the given wood species. This may happen only as a consequence of clogging due to surface deformation by the grits.

Finally, we recall that the direction of sanding in relation to the grains may change the gloss ratios in parallel and perpendicular to the grains (see in Sect. 3.5).

4.9 Effect of Tool Wear on Surface Roughness

As explained in Sect. 4.3 in detail, the action of forces in the chip separation zone is associated with deformations and crushing effects in a thin layer just below the knife-edge which appears as a roughness. The surface roughness is a combination of machining and the inherent anatomical roughness.

Woodworking tools generally operate at high speed, and their working surfaces are subjected to high pressures and friction forces. Consequently, woodworking tools operate under heavy mechanical and thermal loads which are the main reason for tool wear.

The major tool wear mechanisms are *abrasion* as a mechanical wear process, *electrochemical* wear which is attributed to the extractives of timber materials and *high temperature corrosion* due to the extremely high surface temperature of knives (Csanády and Magoss 2013). The abrasion mechanism is always present and it is associated with hot corrosion under high-performance operating conditions.

The dynamics of edge wear depends on the knife material, on the mechanical properties of wood and on the operational parameters. The most important operational parameter is the cutting speed which fundamentally determines how much friction power turns into heat and the surface temperature of the edge. It is not surprising that the Taylor tool-life equation contains the cutting speed v in the following form:

$$v \cdot T^n = \text{const.}$$

where T is the working time of a tool between two sharpening. The exponent n has typical values around 0.5 (Csanády and Magoss 2013).

During the wear process, the profile of the edge maintains its almost semicircle form with increasing radius. This increasing edge radius, which means an increasing loading surface, is responsible for deeper crushing effects and higher core depth R_k.

The crushing effect of a blunt tool is shown in Fig. 4.46 showing the roughness profiles for an Oak sample (Magoss and Sitkei 2000). Using a sharp edge, the roughness profile is mainly determined by the structure of wood and only small peaks are seen on the upper surface due to machining. A blunt edge deforms the upper layer and creates a wavy surface and it differs fundamentally from the profile made by the sharp edge.

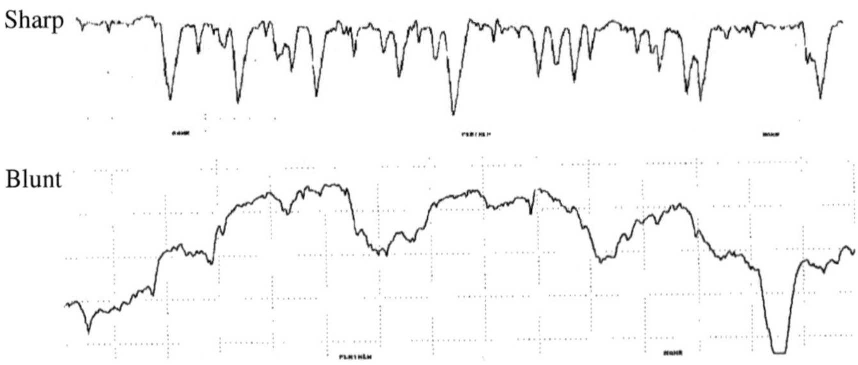

Fig. 4.46 Roughness profiles for an Oak sample machined with sharp and blunt tool

1

R_a=3.95 μm
R_z=37.11 μm
R_{pk}=5.63 μm
R_k=10.22 μm
R_{vk}=10.96 μm

2

R_a=7.64 μm
R_z=55.10 μm
R_{pk}=7.71 μm
R_k=23.65 μm
R_{vk}=15.81 μm

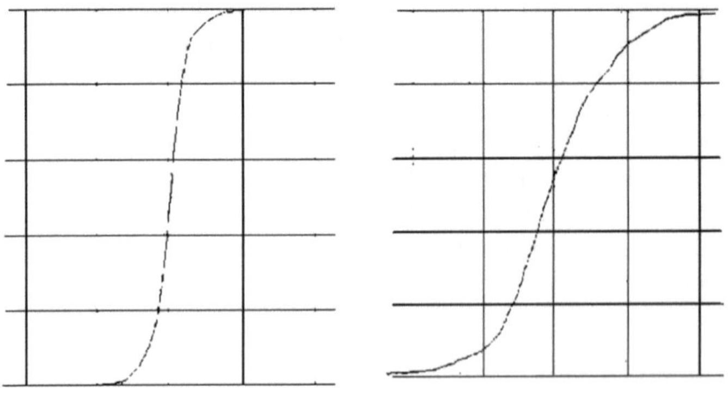

Fig. 4.47 Abbott distribution curves for Beechwood milled with sharp tool (*1*) and after a feed distance of 1800 m (*2*)

The effect of edge blunting can also be followed in the Abbott distribution curve. Figure 4.47 demonstrates curves for a Beech sample milled with a sharp tool (1) and after a feed distance of 1800 m (2). It is clearly seen that the thickness of the deformed zone increased several times, causing surface instability.

The photographs taken by an electron microscope are another kind of experimental evidence clearly demonstrating the compaction of several cell layers under blunt tool edges (Fisher and Schuster 1993; Gottlöber 2014). The compacted layer loses its original structure and becomes very sensitive to mechanical effects and moisture.

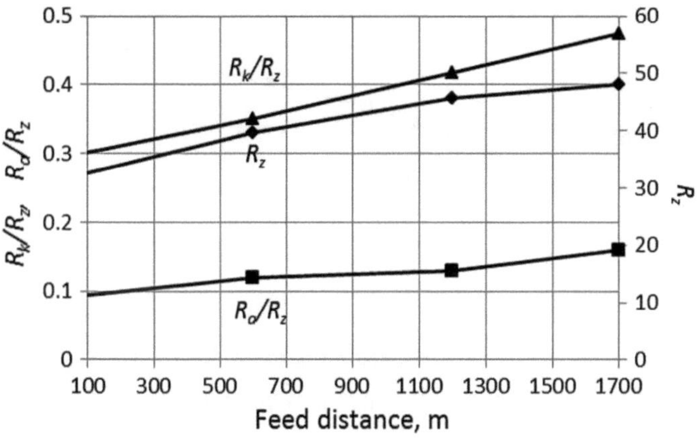

Fig. 4.48 Variation of R_z and its ratios with R_k and R_a as a function of feed distance for Beechwood

The change of edge radius as a function of working time generally shows a parabolic function, while the relation between edge radius and roughness is more or less a linear function. Figure 4.48 represents the averaged roughness height R_z as a function of feed distance using Beech samples. The figure also shows ratios of different roughness parameters (Magoss 2015). All the curves have the form of a parabolic function. The increase of the R_k/R_z ratio indicates that the core depth R_k grows more rapidly than R_z as a function of feed distance or working time. The same is true for the R_a/R_z ratio which indicates that the variation of R_z alone does not fully characterise the surface deterioration due to edge wear. Although R_a is an integrated value, it responses here more sensitive to structural changes in the surface than R_z does.

It is interesting to note that the R_a/R_z ratio for sharp tools varies in a very narrow range for a wide variety of wood species (from conifers to Oak) and it is averaged to 0.11.

The most sensitive and useful roughness parameter is the core depth R_k which uniquely indicates the change in the edge radius due to wear. Fortunately, the R_k roughness parameter may be regarded as almost constant (around 10 μm) for a wide variety of wood species using sharp tools, except the very heavy wood species. In the course of edge wear, this value will be at least doubled as shown in Fig. 4.49 for Beechwood.

Quite similar results were obtained for Oak, Larch and Scots pine (Magoss and Sitkei 2001). Due to the parabolic change in time, the R_k parameter is suitable to follow the edge wear process in its initial time. The reduced peak height R_{pk} steadily increases as a function of wear, but due to its smaller values and greater scattering, its prediction accuracy is inferior compared to R_k. Furthermore, R_{pk} is much less connected physically to phenomena taking place in the core zone.

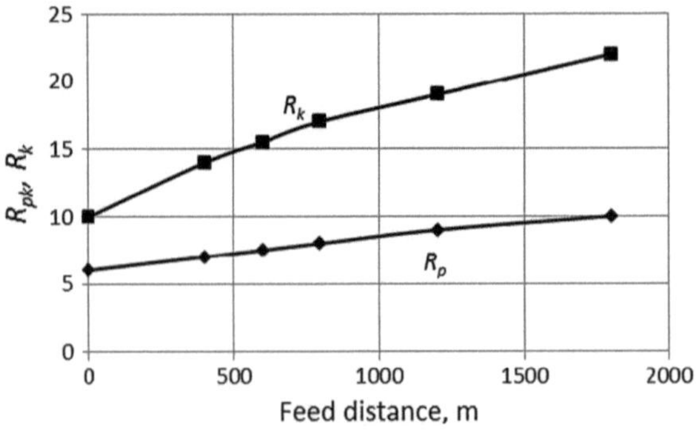

Fig. 4.49 Diagram of R_k and R_{pk} versus feed distance for Beechwood

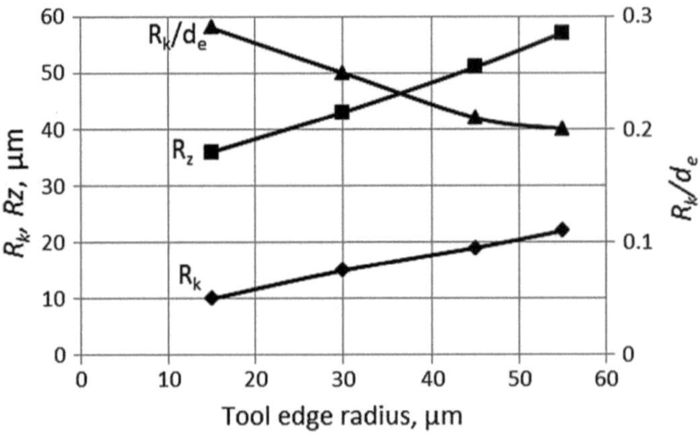

Fig. 4.50 Variation of roughness parameters as a function of tool edge radius. Beech sample

Choosing the tool edge radius as a variable, the R_z and R_k parameters vary linearly as a function of edge radius as depicted in Fig. 4.50. It is interesting to represent the R_k/d_e ratio (d_e is the edge diameter) which should be characteristic for the effect of edge compaction in the core thickness. As shown in Sect. 4.8, this ratio for different grit sizes with good approximation remained constant for all wood species. Figure 4.50 shows that this ratio for knife-edge slightly increases towards sharper edges. Keeping in mind Fig. 4.9 in Sect. 4.3 (an infinitely long cylinder exerts a given pressure twice as deep as a sphere), over edge radii of 50 μm, the knife-edge behaves quite similarly to the spherical grits. The slightly larger R_k/d_e ratio for sharp tools may be explained in the following manner.

From the general theory of woodcutting, it is well known that the radial force component for sharp tools changes its sign, from negative to positive sign (Csanády and Magoss 2013). Due to tension forces in the separation zone just before the edge, some fibres may be torn out from the developing surface and which slightly increases the core depth. With higher edge radii, the radial component is always compressive which does not create any additional roughness.

Earlier experiments have shown that different wood species have nearly the same variation in R_k values as a function of tool edge radius as given in Fig. 4.50 (Magoss and Sitkei 2001). Therefore, the variation of R_k independently of wood species may serve as a useful indicator of tool wear.

In conclusion, tool wear is a major factor considerably influencing the surface quality obtainable by a given machining process. Therefore, the maintenance of tools is of great importance, and the sharpening cycle of tools is determined by surface quality requirements.

4.10 Effect of Wetting on Surface Roughness

Due to their capillary-porous nature, wood materials always contain some water. Since a wood surface is in interaction with the surrounding air, it tends to be in equilibrium with it. In many cases, the wood surface is wetted for various purposes, e.g. aqueous clear coating, and a part of the liquid infiltrates into the wood and the other part evaporates into the air.

The movement of moisture in wood can be described by a diffusion-type differential equation (as introduced in details in Sect. 2.4). The characteristic feature of moisture movement is its unsteady state with moisture gradient (see Fig. 2.13 in Sect. 2.4.2). As a consequence, swelling and shrinkage stresses develop with subsequent residual deformations. During swelling and shrinkage, the surface and a thin layer below are in movement, causing surface distortion (raised grain) and also changes in surface roughness.

During its movement, a wood surface may have another interesting feature. Due to the movement of the surface and to permanent deformations, microcracks may occur, accelerating the diffusion and facilitating the infiltration of liquids, possibly also the aqueous coatings, into the upper layer of wood. Furthermore, the movement may occasionally open voids which would otherwise be closed to free infiltration. This phenomenon may influence not only the surface roughness but also the adhesion properties of coating films on wood surfaces depending on the surfacing process (planing or sanding, etc.) and wood anatomy (Richter et al. 1995).

Surface roughness affects the wetting properties of wood surfaces. Therefore, it is interesting to know the effect of wetting on the surface roughness modification for different wood species and surfacing manufactures.

No systematic research has been found on the subject, and therefore, a detailed experimental research programme was undertaken to clarify the fundamental relationships concerning different wood species (Spruce, Scots pine, Larch, Oak, Black locust, Beech and Poplar) and surfacing techniques (planing, sanding, milling, finishing and thermosmoothing). Distilled water was used for wetting, and the initial thickness of the film was around 100 and 180 μm. In order to follow the changes in the surface just after the wetting, a contactless 3D laser measuring unit

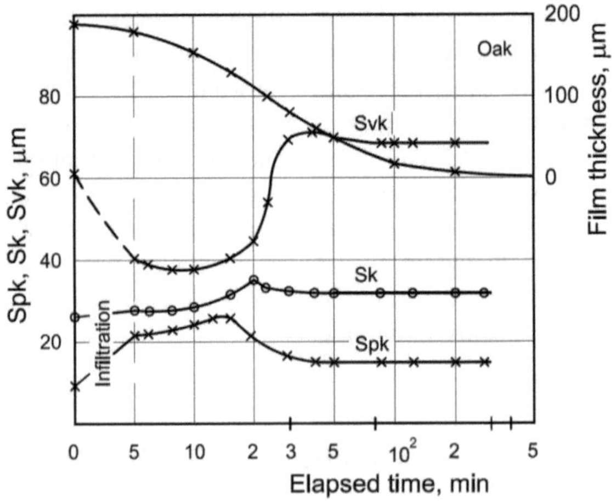

Fig. 4.51 Variation of roughness parameters during the wetting process. Oak planed. Initial film thickness is 180 μm

was used. All important 3D parameters, including the amplitude parameters such as S_q (RMS), skewness and kurtosis of surface height distribution, were recorded and processed (Molnár 2015).

In order to get reliable average values, measurements were made on 10 different spots of each sample one after the other at 1-min intervals. The measurement started 5 min after wetting, and therefore, average values were given only after 15 min. Generally, after 1 h, already no changes were registered, although the measurements were continued for 2 h.

The typical course of a wetting process for planed Oak sample with large vessels is illustrated in Fig. 4.51 showing the momentary values of S_{pk}, S_k and S_{vk} roughness parameters.

After 5 min of infiltration, the record began with significant deviations compared to the initial values. The R_{pk} and R_k values increased during the infiltration time. On the contrary, the R_{vk} value rapidly decreased (with all probability, it is due to swelling) to a minimum one and after some 15 min rapidly increased to values higher than the initial one. All three parameters have a local maximum culminating in subsequent elapsed times according to their vertical positions. The local maximum of the lower S_{vk} layer coincides with the half-time of moisture evaporation (30 min). After the local maximum point, all three parameters proceed asymptotically to their final values which are higher than their initial values. After 1 h, no change was observed.

The strict variation of all three roughness parameters indicates intensive movements and deformations in the upper surface layer with permanent deformations.

A somewhat different behaviour was recorded for Scots pine samples without vessels. After wetting, all three roughness parameters increased to a maximum and after culminating decreased to their final values. Until S_{pk} and S_{vk} are moderately increased, S_k shows an unexpectedly high increase.

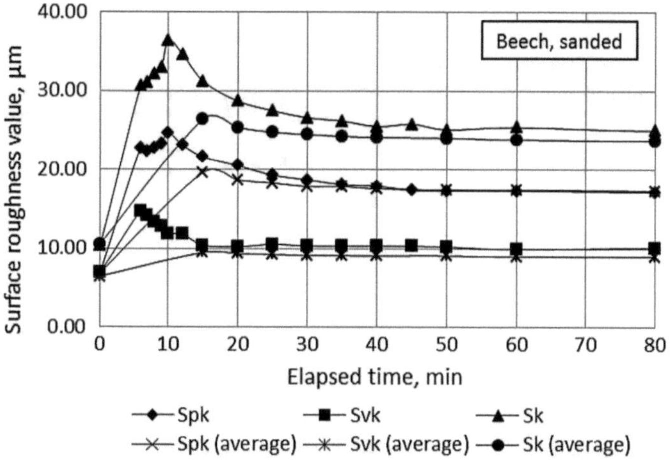

Fig. 4.52 Variation of momentary and average roughness parameters after wetting as a function of time. Beech sample, sanded. Initial film thickness is 100 μm

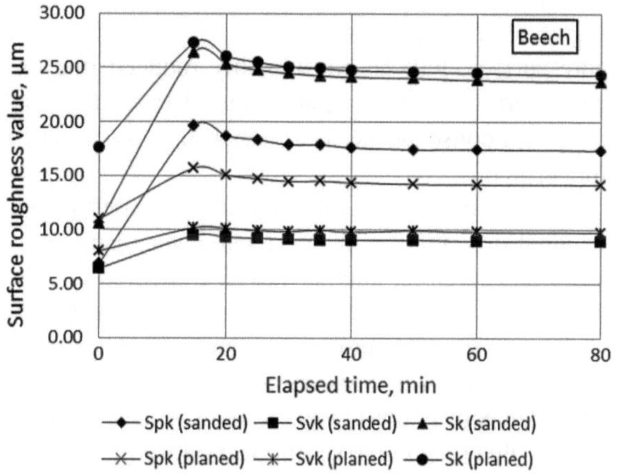

Fig. 4.53 Variation of averaged roughness parameters after wetting as a function of time. Beech samples planed and sanded

Figure 4.52 shows similar results for a sanded Beech sample, giving both the momentary and average values. After half an hour, the momentary and average values measured on ten different spots approach each other. Again, the core thickness S_k is highly increased, confirming the movement and residual deformation of the upper layer due to swelling and shrinkage. The smallest changes occur in the deeper valleys as indicated by the S_{vk} parameter.

As a comparison, Fig. 4.53 represents average roughness values for the same Beech sample but with two differently manufactured surfaces. Considerable difference occurs in the uppermost layer affecting the S_{pk} values. The sanded surface suffers more residual deformations, increasing the averaging height of the protruding peaks above the core surface. The core thickness S_k has also increased

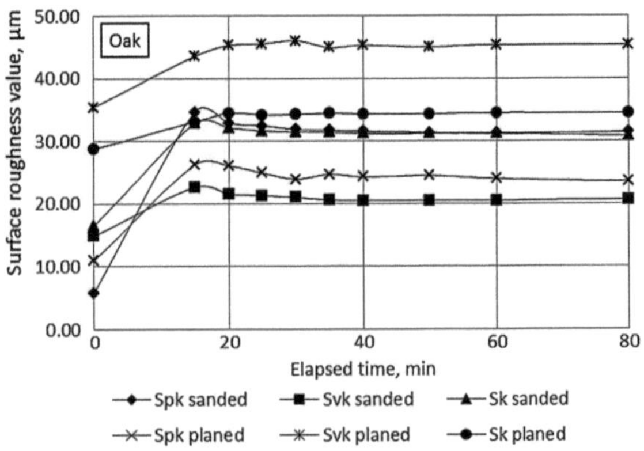

Fig. 4.54 Averaged roughness parameters after wetting as a function of time. Oak sample, planed and sanded

considerably, but taking into account their initial values, the increase is not the same for the planed and sanded surfaces.

Hardwood species, such as Oak, may show a somewhat different picture. The momentary peak values in the first 6–10 min are much less distinct. Due to the large vessels, the clogging effect of sanding prevails in a greater extent as shown in Fig. 4.54. The parameter S_{vk}, the averaging height of protruding valleys below the core surface, will be considerably reduced due to sanding. The uppermost layer characterised by the S_{pk} parameter, after sanding, suffers more residual deformation and, as a consequence, has higher S_{pk} values. The core thickness S_k also became higher here.

The analysis of a vast amount of experimental results has revealed that the residual deformation in the upper layer restructures the distribution of peaks and valleys. Therefore, another possible method to follow the changes due to wetting may be the use of distribution parameters such S_q, S_{ku} and S_{sk}.

Several researchers from the metal industry have suggested that the use of both S_a and S_q (RMS) is unnecessary because these parameters do not differ much. But this statement is not always true, especially concerning a wood surface. It can be shown theoretically (see in Sect. 4.4) that the S_q/S_a (or R_q/R_a) ratio uniquely depends on the shape of protruding peaks and valleys and may vary in a broad range, at least between 1.1 and 1.9. Therefore, the S_q/S_a ratio may be an important indicator to characterise roughness distribution.

The functional relationships of $S_{ku} = f(S_{sk})$ or $S_{sk} = f(S_q/S_a)$ may be used to examine surface profile distribution. As is well known, for a surface with symmetric height distribution, the skewness is zero. An asymmetric height distribution may be negative or positive depending on the position of its maximum.

Kurtosis S_{ku} is generally used with skewness to characterise the shape of height distribution. For a Gaussian surface, the kurtosis has a value of 3. A sharp, centrally distributed height distribution has values larger than 3. On the contrary, a well-spread distribution has kurtosis values less than 3.

Figure 4.55 shows the movement of the profile height distribution due to wetting for Spruce samples with different surfacing. There is a general tendency that the

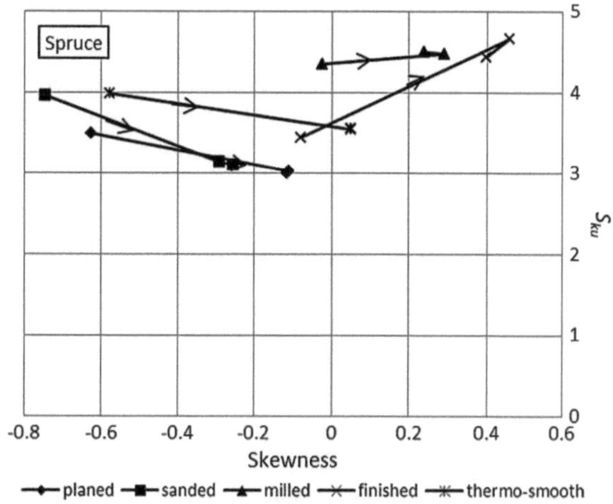

Fig. 4.55 Variation of skewness of a Spruce sample with differently manufactured surfaces

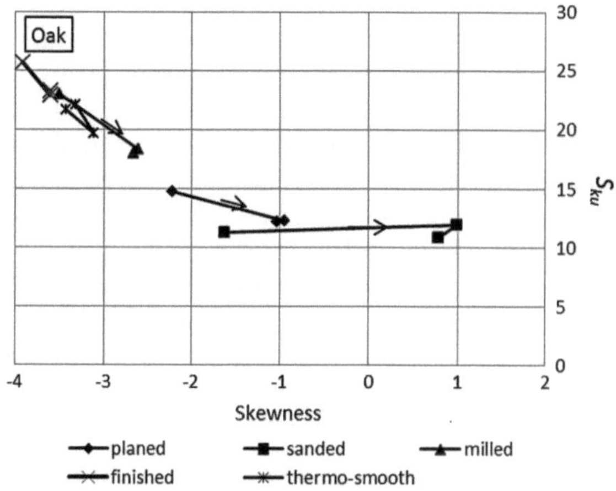

Fig. 4.56 Skewness variation of Oak samples due to wetting for differently manufactured surfaces

initial negative skewness after wetting moves towards positive skewness. Another observation is that for all softwood species (Spruce, Larch and Scots pine), the sanded surface has always the highest negative skewness in its initial condition. Due to wetting, this skewness is getting near to zero. The finished surface has started from around zero skewness and was getting to the highest positive skewness up to 0.3–0.5.

The hardwood species (Oak, Black locust and Beech) have another behaviour concerning their surfacing manufacture. Figure 4.56 demonstrates skewness variations after wetting for Oak samples. The initial skewness is highly negative with values up to −4 (finished surface). The sanded surface has here, however, the smallest negative skewness, and after wetting, it is getting into the positive skewness range. The length of vectors indicating the variation of skewness is a visual measure for quick estimation. Similar results were obtained for Beech and Black locust.

Concerning all wood species tested, the sanded surface has suffered the greatest variation in its skewness and the variation always occurred towards the positive range. This observation shows that movements and residual deformations are also the highest in the sanded surface independently of wood species. This movement due to wetting may be a new and further explanation to earlier observations that surface sanding provides the best conditions for coating and paint applications and requires the lowest quantity of paint for good coverage (Richter et al. 1995).

Adhesion of coating to the surface will be more favourable if it takes place in a three-dimensional interface providing a transition layer between wood and coating. Such an interface is highly promoted by diffusion which, in turn, is accelerated by the movements in the surface.

The S_q/S_a ratio for conifers is grouped around 1.3 and for Beech and Black locust between 1.35 and 1.45, respectively, while the widest range was obtained for Oak up to 1.9.

Figure 4.57 shows the skewness and S_q/S_a relationship for Oak samples with differently manufactured surfaces. A finisher with large oblique angle and sharp edge cuts a very smooth surface with a highly asymmetric height distribution (similar to a plateau-honed metal surface). A sanded surface has the least asymmetric initial distribution which moves into the positive skewness range after wetting. All other surfaces remained in the negative skewness range after wetting. The smallest movements occur in finished and thermosmoothed surfaces which provide an inferior surface for coating and painting purposes.

The deformation in the surface layer depends on the applied water thickness and the evaporation rate. Figures 4.51 and 4.52 show an initial water thickness of 180 μm is associated with changes in the R_{vk} layer, while that of 100 μm is perceptible only in the R_k and R_{pk} layers.

The rate of evaporation may also considerably change the picture.

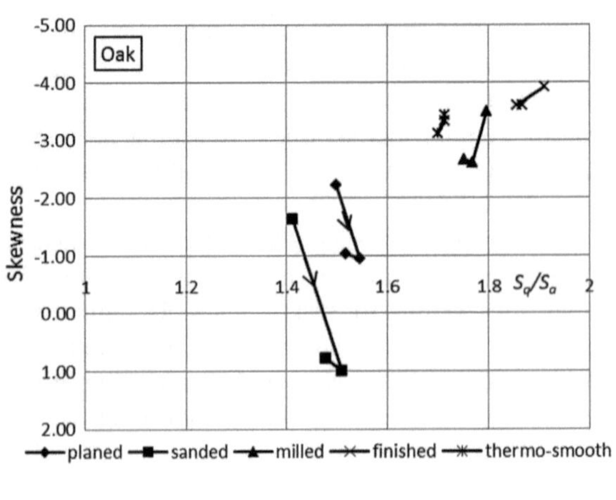

Fig. 4.57 Interrelation of skewness and S_q/S_a ratio for Oak samples with different surfacing manufactures

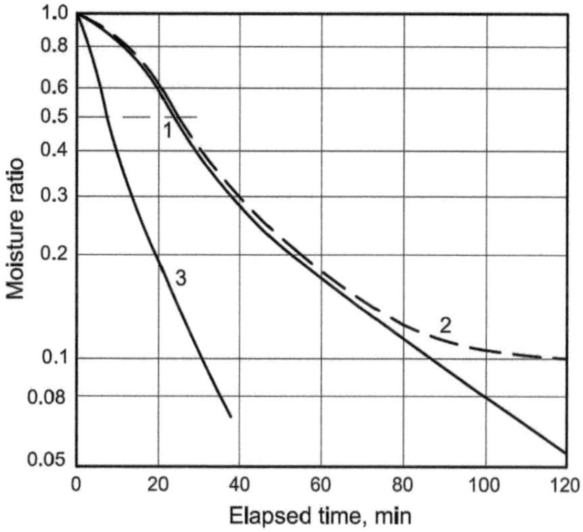

Fig. 4.58 Evaporation of water on an Oak surface. *1*—room temperature, 100 μm, *2*—room temperature, 180 μm, *3*—hot air (60–65 °C), 100 μm

Evaporation at room temperatures has half-times of 23 and 25 min for 100 and 180 μm thicknesses, respectively. With higher film thicknesses, the rate of evaporation suddenly decreases as the surface dries out and the moisture is conducted back to the surface. Using hot air drying between 60 and 65 °C, the half-time was reduced to 8 min for 100 μm film thickness, as shown in Fig. 4.58. The fast evaporation hinders the penetration of water into the deeper layers, and the increase in roughness parameters was much reduced compared to the non-wetting case (Fig. 4.59).

Summarising the experimental results, the effect of wetting on surface roughness may be quite different depending on conditions. Water penetrating into the upper surface layer always causes greater and smaller increase in the roughness parameters due to swelling and shrinkage, and permanent deformations. This phenomenon may improve the strength of gluing due to the movements in the surface layer promoting a

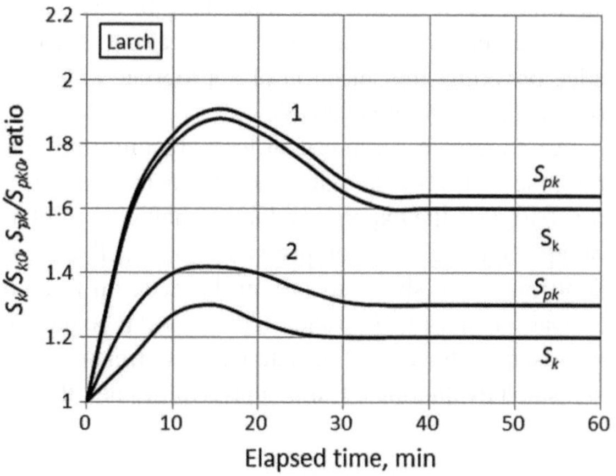

Fig. 4.59 Roughness parameter ratios related to their initial values for larch sample, planed. *1*—drying at room temperature, *2*—dried with hot air

more uniform capillary coating penetration. The depth of coating penetration depends on the viscosity of coating. For example, due to its higher viscosity, a pigmented coating penetrates less deeply into a surface (de Meijer et al. 2001).

4.11 Scattering of Roughness Data

The major part of the resultant roughness originates from the anatomical structure of wood. Cavities in the wood material will be cut at different angles and positions in relation to the surface, leaving valleys of various shapes and sizes on the surface. That is a random process causing the scattering of the measurement results, and it may be treated with statistical methods.

The main components determining the scattering of roughness are as follows:

- component due to the brittle fracture of wood during cutting;
- component due to structural differences in earlywood and latewood;
- component due to the random position of the cutting plane to vessels and tracheids.

Seasonal changes in the earlywood-and-latewood ratio may cause additional scattering. The high variability of structural properties within a given species further complicates the problem. The most important relationships of structural variations were treated in Sect. 4.2 (see Figs. 4.3, 4.5 and 4.6).

Machining also contributes to the scattering of roughness. The primary reason for the roughness caused by machining is the brittle fracture of wood cells and its low tensile strength perpendicular to the grain. The brittle fracture cannot eliminate, but it can be limited to a low volume. The most effective method is high-speed cutting with the smallest material contact possible (using a sharp edge).

To eliminate the negative effect of low tensile strength perpendicular to the grain, it is important to generate a compressive load in the immediate vicinity of the cutting edge. This can be done using cutting angles in the range of 65–70° or rake angles of 20–25°. The edge machining of boards requires special attention because they are very inclined to the edge breaking due to tensile loads. In order to avoid edge breaking, the selection of appropriate kinematic conditions is very important (Csanády and Magoss 2013).

An excessive compression load crushing the material can also cause additional roughness. The method of "smooth machining" has been known for a long time and it is based on the experience that a small chip thickness raises smaller forces. Using oblique cutting, the compressive loads will be considerably reduced due to the contribution of shear stresses in the material fracture (Sitkei 1997).

Data scattering as a random process can be handled with statistical methods. It is necessary to have a set of data which correspond to normal distribution for reliable statistical evaluations. Using a probability net, the plot of experimental results gives

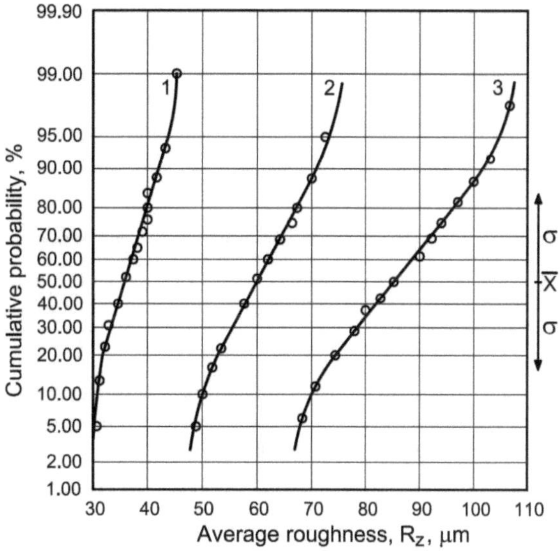

Fig. 4.60 Measurement data distribution for Scots pine (*1*), Ash (*2*) and Oak (*3*) samples

Table 4.3 Relative scattering of R_z values

Wood species	$\sigma/\overline{X}$
Scots pine	0.13
Larch	0.14
Poplar	0.125
Beech	0.12
Black locust	0.135
Ash	0.14
Oak	0.14

Table 4.4 Relative scattering for Spruce and Oak

Wood species	R_{max}	R_z	R_a	R_{pk}	R_k	R_{vk}
Spruce	0.15	0.09	0.08	0.11	0.08	0.1
Oak	0.2	0.1	0.095	0.2	0.11	0.15

a straight line, provided that the set of data fulfil the above requirement. At least 60–100 measurement data are required to achieve this aim.

Figure 4.60 shows typical measurement results on milled samples of Scots pine, Ash and Oak (Magoss 2008). The curves are straight with some deviations at their ends due to their incomplete distribution which has a simple physical explanation: infinitely low and high values do not occur in real distributions. Using the figure, the appropriate values of the average roughness height and its standard deviation can be read. The standard deviation related to the average value gives the relative scattering which is summarised in Table 4.3 for different wood species.

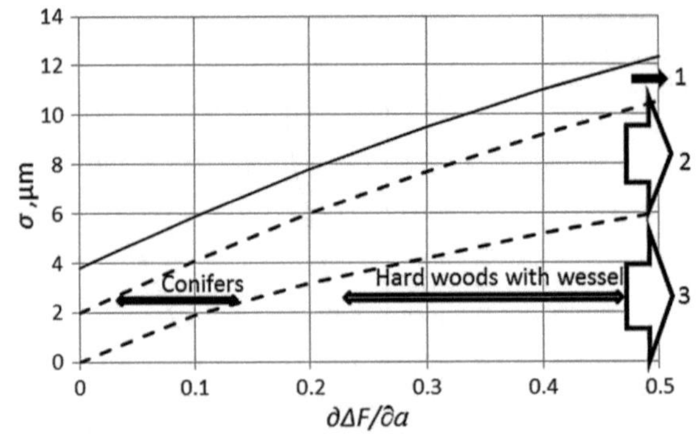

Fig. 4.61 Standard deviation of the roughness parameter R_z as a function of the derivative $\partial \Delta F / \partial a$ for different wood species. *1*—roughness component due to machining, *2*—component due to random position of cutting plane to vessels and tracheids, *3*—component due to structural differences in the earlywood and latewood

The relative scattering of the different wood species falls in a surprisingly narrow range, between 0.12 and 0.14. This circumstance facilitates the rough estimation of the scattering based on the average values. Further experiments have revealed that the different roughness parameters may have dissimilar relative scattering. Performing 30 measurements on samples side by side showed relative scattering as given in Table 4.4. In this experiments, Spruce systematically had lower scattering compared to Oak. R_{max} and R_{pk} are the most sensitive to scattering, while R_a and R_k are less prone to scattering.

Keeping in mind Fig. 4.5 in Sect. 4.2, the standard deviation referred to R_z can be represented as a function of the derivative with respect to the earlywood portion "*a*" of the structure number. This relationship, using the data for different wood species, is given in Fig. 4.59 (Magoss 2008). The resultant standard deviation naturally contains all three components listed above.

The reliable separation of these components seems to be a difficult task, and only a rough estimate can be imagined. Knowing the participation of the roughness component due to machining (see Fig. 4.33) in the resultant one, an estimate may be made with respect to its contribution to the resultant standard deviation. The separation of the other two components of the standard deviation is more difficult, and it may only be supposed that they may have the same magnitude. With these assumptions, a possible separation is given in Fig. 4.61.

Conifers fall in a narrow range, and their standard deviations vary between 4 and 6 μm. Beech and Poplar are close to conifers with a range between 5.5 and 6.5 μm. The hardwoods with large vessels occupy a broader range between 8 and 12 μm. All these figures refer to the averaged roughness height R_z.

The measurement of roughness parameters is always associated with a considerable scattering inherent mainly in the internal structure of wood. It should also be added that the measurements are not free from errors. Roughness parameters are imperfect statistical representations due to the very complicated geometrical structures, and therefore, they are more prone to scatter as a consequence of random sampling (Thomas 1981).

References

Abbott, E., Firestone, F.: Specifying surface quality. Mech. Eng. **55**, 569–572 (1933)

Csanády, E., Magoss, E.: Mechanics of Wood Machining. Springer, Berlin (2013)

Dobler, K.: Der freie Schnitt beim Mähen von Halmgut. Hohenheimer Arbeiten No. 62, 1972

Dong, W.P., Sullivan, P.J., and Stout, K.J.: Comprehensive study of parameters characterising three dimensional surface topography. Wear **178**, 29–43 (1994)

de Meijer, M., Thurich, K., Militz, H.: Quantitative measurements of capillary coating penetration in relation to wood and coating properties, pp. 35–45. Holz als Roh- und Werkstoff (2001)

Dobrint, P.: Optimiertes Schleifen von MDF und Spanplatten. Tagungsband, 8. Holztechnisches Koll, pp. 125–136. Braunschweig (1991)

Fischer, R., Schuster, C.: Qualitätsentstehung spannend erzeugter Holzoberflächen. Mitteilung aus dem Institute für Holztechnik der TU Dresden (1993)

Gottlöber, C.: Zerspanung von Holz und Holzwerkstoffen: Grundlagen – Systematik – Modellierung – Prozessgestaltung. Fachbuchverlag Leipzig im Carl Hanser Verlag, München (2014). ISBN 978-3-446-44058-6, E-Book-ISBN 978-3-446-44003-6

Кряжев Н.А.: Цилиндрическое и коническое фрезерование древесины. М., 1963. (Cylindrical and cone –shaped tools for wood milling.)

Lang, M.: Untersuchungen zur Entwicklung eines kombinierten Plan- und Umfangfräsverfahrens mit kegelstumpfförmigen Werkzeugen. Ph.D. Dissertation, Dresden-Sopron, 1989

Lehmann, B., Hutton, S.: The kinematics of washboarding of bandsaws and circular saws. In: Proceedings of 13th IWMS Vancouver, pp. 205–216, 1997

Magoss, E., Sitkei, G.: Optimum surface roughness of solid woods. In: Proceedings of 16st IWMS Conference, Matsue, pp. 366–371, 2003

Magoss, E., Sitkei, G.: Strukturbedingte Rauheit von mechanisch bearbeiteten Holzoberfächen, pp. 231–239. Möbeltage in Dresden, Tagungsbericht (2000)

Magoss, E.: Insight into the mechanism of wood sanding. Progress Rep. No. 3, DWE University of West Hungary, Sopron (2015)

Molnár, Zs.: Végmegmunkált természetes faanyagok felületi stabilitása nedvesítéskor (Stability of machined wood surfaces at wetting). Ph.D. Dissertation, Sopron (2015)

Magoss, E., Sitkei, G.: Fundamental relationships of wood surface roughness at milling operations. In: Proceedings of 15th IWMS Conference Los Angeles, pp. 437–446, 2001

Magoss, E.: General regularities of the surface roughness sanding solid woods. In: Proceedings of 21st IWMS Conference, pp. 325–332, 2013

Magoss, E.: General Regularities of Wood Surface Roughness. Progress Rep. No. 1, DWE University of West-Hungary, Sopron (2008)

Okay, R., Yokochi, H., Kimura, S.: What is relationship between tooth passage frequency and natural frequency of the bandsaw when washboarding induced. In: Proceedings of 12th IWMS Kyoto, pp. 367–380 (1995)

Orlowski, K., Wasielewski, R.: Washboarding during cutting on frame sawing machines. In: Proceedings of 15th IWMS Los Angeles, pp. 219–228, 2001

Richter, K., Feist, W.C., Knaebe, M.T.: The effect of surface roughness on the performance of finishes. Forest Prod. J. **45**, 91–97 (1995)

Scholz, F., Ratnasingam, J.: Optimization of sanding process. In: Proceedings of 17th IWMS Conference Rosenheim, pp. 422–429, 2005
Siklinka, M., Ockajova, A.: The study of selected parameters on wood sanding. In: Proceedings of 15th IWMS Conference Los Angeles, pp. 485–490, 2001
Sitkei, G.: On the mechanics of oblique cutting of wood. In: Proceedings of 13th IWMS Conference, pp. 469–476, 1997
Sitkei, G.: Theorie des Spanens von Holz. Fortschrittbericht No. 1 Acta Fac. Ligniensis Sopron (1990)
Sitkei, G.: Zur Beurteilung der Farbtöne von Holzarten, p. 32. Fortschrittbericht No. 2. Institute für Holzbearbeitung. Universität von Sopron (2013)
Thomas, T.: Characterization of surface roughness. Prec. Eng. **3**, 97–103 (1981)
Tröger, J., Lang, M.: Hobelfräsen mit verminderten Wellenschlag, pp. 43–49. HOB, (1990)
Timoshenko, S., Goodier, J.: Theory of Elasticity. New York (1951)

Appendix 1

Selected tree species having distinct colour hues.

Yellow-Coloured Species

Damar minyak (*Agathis borneensis*)
Huon pine (*Lagarostrobos franklinii*)
Jelutong (*Dyera costulata*)
Ramin (*Gonystylus bancanus*)
Sandalwood (*Santalum album*)
Santinwood (*Chloroxylon swietenia*)
Terap (*Artocarpus elasticus*)
Yellow Meranti (*Shorea resina-nigra*)
Yellowwood (*Podocarpus falcatus*)

Cream-Coloured Species

Mersawa (*Anisoptera curtisii*)
Rewarewa (*Knightia excelsa*)
Tawa (*Beilschmiedia tawa*)
Tawhai (*Nothofagus menziesii*)
York gum (*Eucalyptus loxophleba*)

Light and Middle Brown Species

Alerce (*Fitzroya cupressoides*)
Angelim pedra (*Hymenolobium modestum*)
Blackwood (*Acacia melanoxylon*)
Kiaat (*Pterocarpus angolensis*)

E. Csanády et al., *Quality of Machined Wood Surfaces*,
DOI 10.1007/978-3-319-22419-0

Kauri (*Agathis australis*)
Kungur (*Pithecellobium* spp.)
Mahogany (*Swietenia mahagoni*)
Matai (*Prumnopitys taxifolia*)
Meranti bakau (*Shorea uliginosa*)
Merpauh (*Swintonia schwenhii*)
Rauli (*Nothofagus alpina*)
Redwood (*Seqoia sempervirens*)
Rimu (*Dacrydium cupressinum*)
Sepetir (*Sindora siamensis*)
Sugi (*Cryptomeria japonica*)
Tasman myrtle (*Nothofagus cunninghamii*)
Totara (*Podocarpus totara*)

Orange-Coloured Species

Rainbow wood (*Caesalpinia sappan*)
Meransi (*Carallia* spp.)

Dark Brown Species

Jarrah (*Eucalyptus marginata*)
Jambire (*Milletia stuhlmannii*)
Kohekohe (*Dysoxylum spectabile*)
Mulga (*Acacia aneura*)
Nyireh (*Xylocarpus moluccensis*)
Puriri (*Vitex lucens*)
Rata (*Metrosideros robusta*)
Siamese rosewood (*Dalbergia cochinchinensis*)
Stinkwood (*Ocotea bullata*)
Sucupira (*Andira micrantha*)
Tamboti (*Spirostachys africana*)
Teak (*Tectona grandis*)

Reddish Brown Species

Balau red (*Shorea guiso/hunstleri*)
Bintangor (*Calophyllum* spp.)
Bitis (*Madhuca utilitis*)
Bubinga (*Guibourtia tessmannii*)

Chamfuti (*Afzelia quanzensis*)
Dark Red Meranti (*Shorea pauciflora*)
Kempas (*Koompassia malaccensis*)
Keruing (*Dipterocarpus alatus*)
Nyatoh (*Madhuca motleyana*)
Red Mahogany (*Khaya antotheca*)
Sapele (*Entandrophragma cylindricum*)
Simpoh (*Dillenia* spp.)
Tualang (*Koompassia excelsa*)
Zambezi teak (*Baikiaea plurijuga*)

Red Colour Species

Binneringie Wattle (*Acacia donaldsonii*)
Bois de rose (*Dalbergia maritima*)
Karri (*Eucalyptus diversicolor*)
Kekatong (*Cynometra inaequifolia*)
Nazareno (*Peltogyne purpurea*)
Pau roxo (*Peltogyne excelsa*)
Red ivory (*Berchemia zeyheri*)
Rengas (*Gluta* spp.)

Blackish and Black Species

Ebony (*Diospyros ebenus*)
Ebony tree (*Euclea pseudebenus*)
Pau Preto (*Dalbergia melonoxylon*)
Wenge (*Milletia laurentii*)

"Ironwood" Species Over 1000 kg/m^3 Density

Ebony (*Diospyros ebenus*)
Ebony tree (*Euclea pseudebenus*)
Balau (*Shorea siamensis*)
Siamese rosewood (*Dalbergia cochinchinensis*)
Cam lien (*Dalbergia bariensis*)
Pau Preto (*Dalbergia melanoxylon*)
Bitis (*Madhuca utilis*)
Sen mat (*Madhuca pasquieri*)
S.A. ironwood (*Olea capensis*)

Kekatong (*Cynometra inaequifolia*)
Keranji (*Dialium* spp.)
Tempenis (*Streblus elongatus*)
Tau mat (*Vatica cinerea*)
Lin xanh (*Erythrophloeum fordii*)
Leadwood (*Combretum imberbe*)
Camel thorn (*Acacia erioloba*)
Wandoo (*Eucalyptus wandoo*)
Black ironwood (*Krugiodendron ferreum*)
Lignum vitae (*Guaiacum sanctum*)
Boxwood (*Buxus colchica*)

Note Under semidesert conditions, the trees generally develop very dense texture with deep colour. Their density may be extremely high, up to 1300 kg/m^3. An exceptionally reach area can be found in Western Australia, where the multitude of "Goldfields" species, mostly eucalyptus and acacias, is simply innumerable. They supply, however, smaller samples.

Appendix 2

Colour characterization of selected wood species
Black locust, *Robinia pseudoacacia*

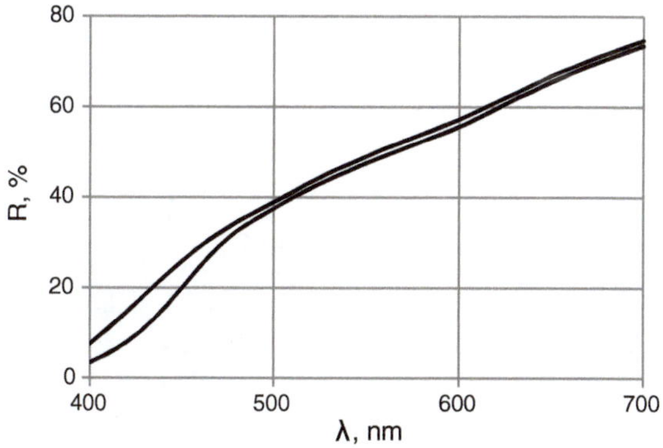

$L_0^* = 75 - 74\ \%$, $R_m = 46.5 - 44\ \%$, $\lambda_0 = 529 - 535$ nm, $h^* = 8.3 - 18$
$K = 1.53 - 1.35$, Colour hue: grey

E. Csanády et al., *Quality of Machined Wood Surfaces*,
DOI 10.1007/978-3-319-22419-0

Scots pine, *Pinus sylvestris*

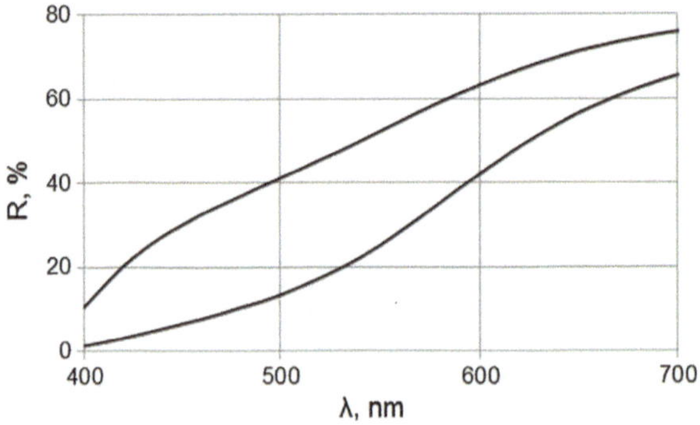

$L_0^* = 77.4/59$ %, $R_m = 50/30$ %, $\lambda_0 = 543/564$ nm, $h^* = 4.65/2.43$
$K = 1.08/0.65$, Colour hue: gr. yellow/y. brown

King Billy pine, *Athrotaxis selaginoides*

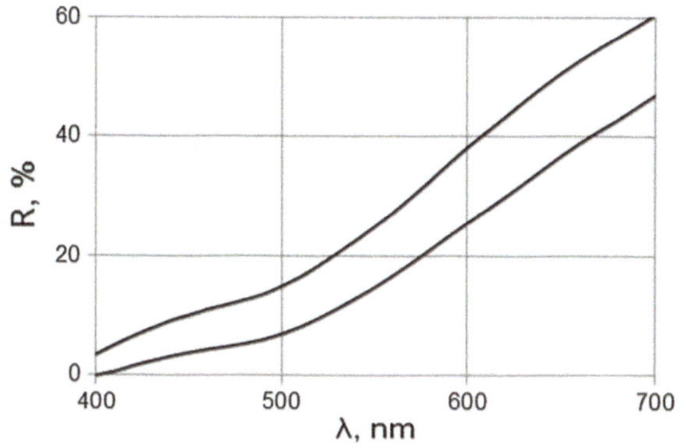

$L_0^* = 58 - 47\ \%$, $R_m = 28.5 - 18.6\,\%$, $\lambda_0 = 565 - 570$ nm, $h^* = 2.1 - 2.17$, $K = 0.6 - 0.5$, Colour hue: y. brown-brown

Sugi (heart), *Cryptomeria japonica*

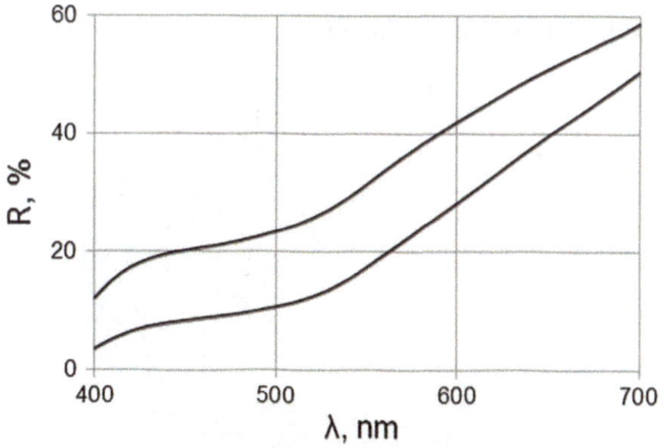

$L_0^* = 64 - 51\ \%$, $R_m = 33.6 - 22\ \%$, $\lambda_0 = 565 - 575$ nm, $h^* = 1.74 - 1.54$ $K = 0.58 - 0.4$, Colour hue: y. brown-brown

Sugi (sap), *Cryptomeria japonica*

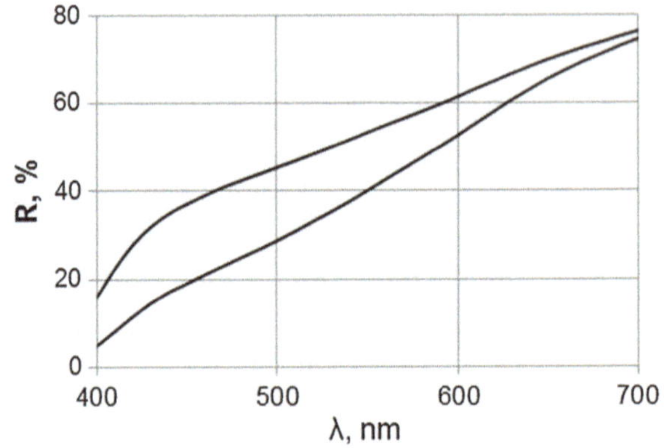

$L_0^* = 78 - 71\ \%$, $R_m = 52 - 41\ \%$, $\lambda_{0=} = 544 - 553$ nm, $h^* = 4.28 - 3.32$, $K = 1.05 - 0.84$, Colour hue: gr. yellow-yellow

Camphor wood, *Cinnamomum camphora*

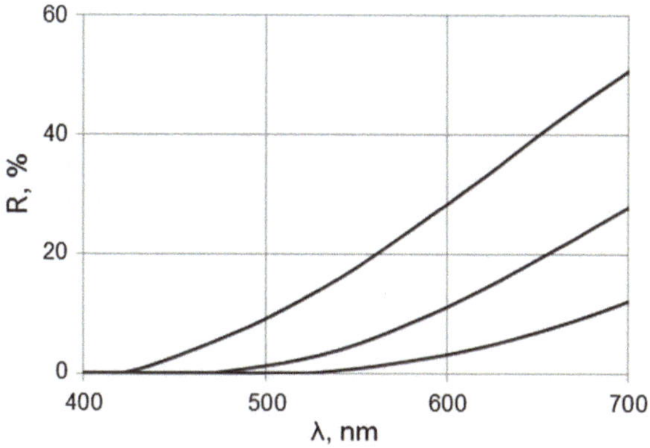

$L_0^* = 50 - 13$ %, $R_m = 20.5 - 2.9$ %, $\lambda_0 = 563 - 593$ nm, $h^* = 3.28 - 0.94$
$K = 0.68 - 0.17$, Colour hue: y. brown-red

Merbau, *Intsia palembanica*

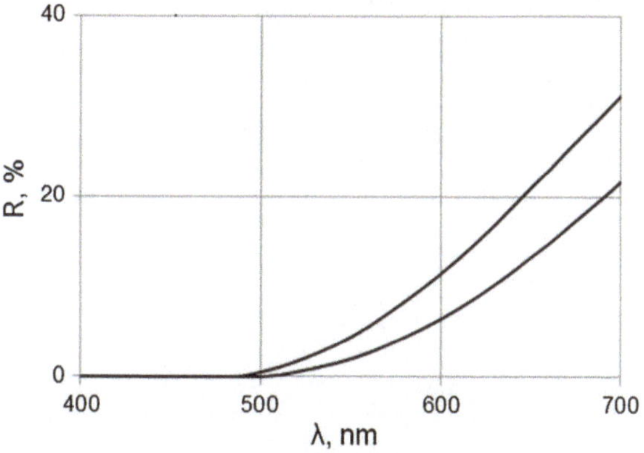

$L_0^* = 29 - 20.5$ %, $R_m = 8.9 - 5.5$ %, $\lambda_0 = 584 - 591$ nm, $h^* = 1.89 - 1.37$
$K = 0.3 - 0.2$, Colour hue: b. red-red

American walnut, *Juglans nigra*

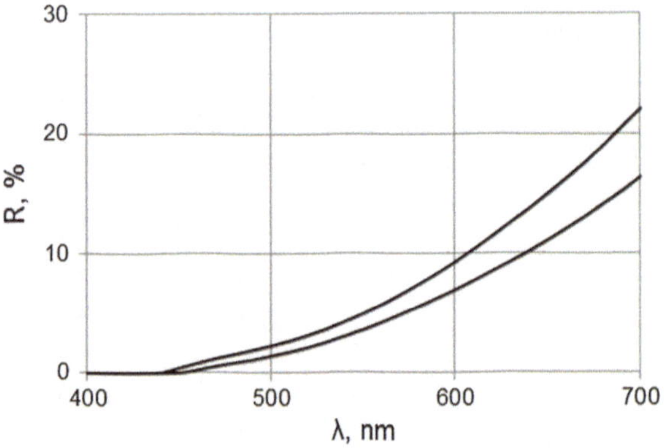

$L_0^* = 28.5 - 24\ \%$, $R_m = 7.15 - 5.22\ \%$, $\lambda_0 = 577 - 576$ nm, $h^* = 2.5$, $K = 0.45 - 0.43$, Colour hue: brown

Sapele, *Entandrophragma cylindricum*

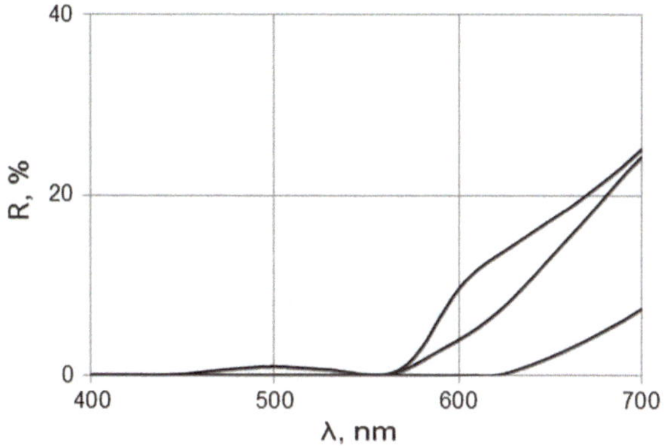

$L_0^* = 20.6 - 0.8$ %, $R_m = 6.7 - 0.95$ %, $\lambda_0 = 595 - 625$ nm, $h^* = 0.88 - 0.25$, Colour hue: red

Wenge, *Milletia laurentii*

butt edge (magnified)

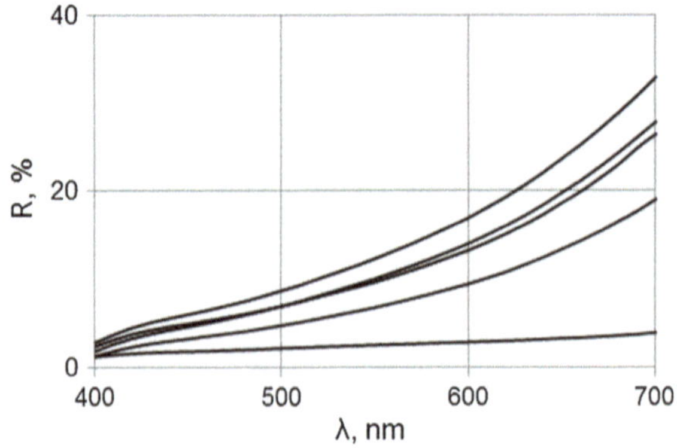

$L_0^* = 42.5 - 18$ %, $R_m = 14.3 - 2.5$ %, $\lambda_0 = 550 - 573$ nm, $h^* = 3.9 - 2.3$, $K = 1.0 - 0.55$, Colour hue: yellow-brown

Doussie, *Afzelia rhomboidea*

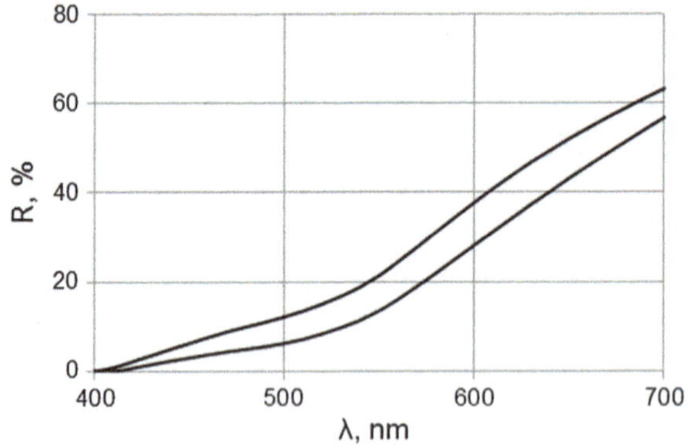

$L_0^* = 56 - 47$ %, $R_m = 27 - 20.6$ %, $\lambda_0 = 570 - 579$ nm, $h^* = 2.17 - 1.86$, $K = 0.5 - 0.36$, Colour hue: brown

Ironwood, *Olea capensis*

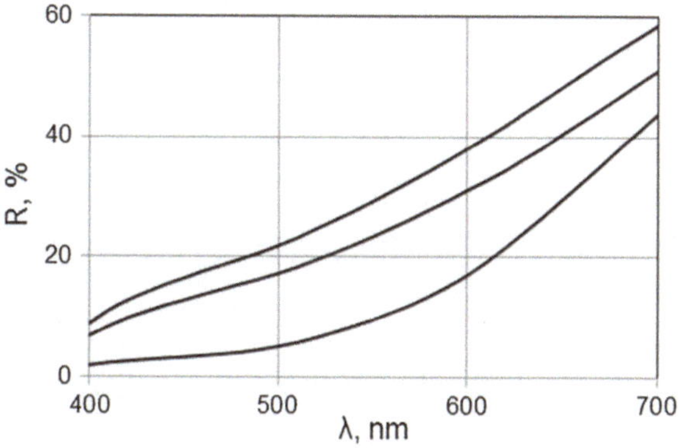

$L_0^* = 53 - 29$ %, $R_m = 22.1 - 6.85$ %, $\lambda_0 = 564 - 577$ nm, $h^* = 2.7 - 2.16$
$K = 0.66 - 0.47$, Colour hue: y. brown-brown

Rewarewa, *Knightia excelsa*

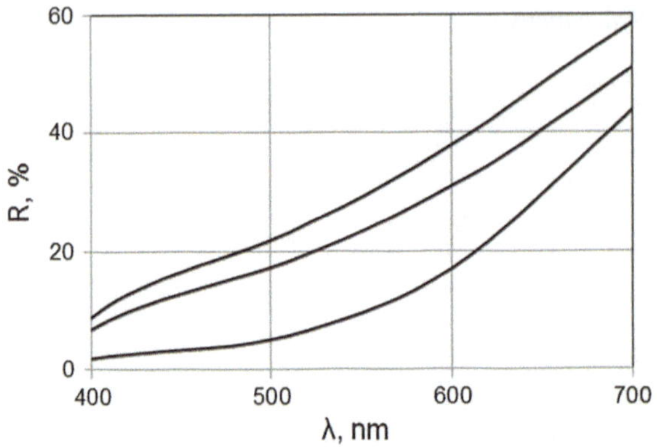

$L_0^* = 62 - 40\ \%$, $R_m = 31.5 - 15\ \%$, $\lambda_0 = 563 - 587$ nm, $h^* = 2.49 - 1.68$, $K = 0.66 - 0.31$, Colour hue: y. brown-br. red

Sucupira, *Andira micrantha*

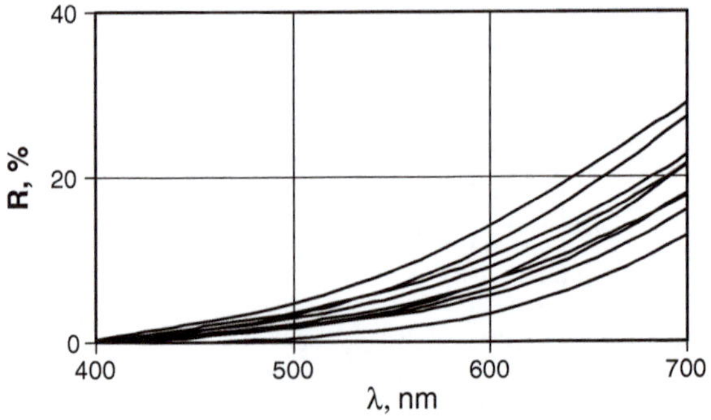

$L_0^* = 36 - 16\ \%$, $R_m = 11 - 3.2$, $\lambda_0 = 573 - 594$ nm, $h^* = 2.3 - 1.6$, $K = 0.5 - 0.24$, Colour hue: brown-red

Angelim pedra, *Hymenolobium modestum*

butt-edge (magnified)

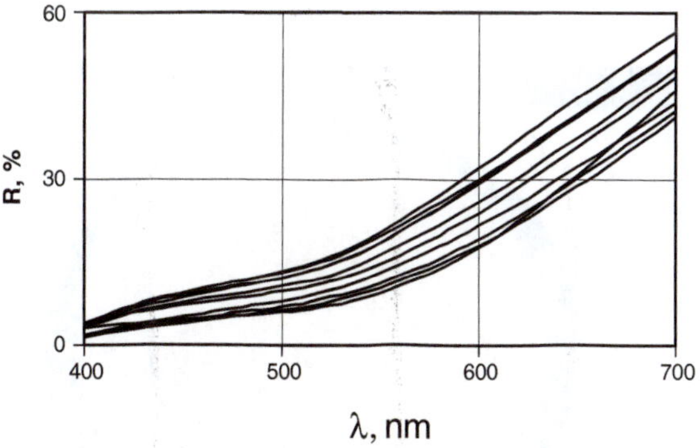

$L_0^* = 54 - 40\ \%$, $R_m = 25 - 15.6\ \%$, $\lambda_0 = 572 - 586$ nm, $h^* = 2.0 - 1.4$, $K = 0.45 - 0.28$, Colour hue: brown-br.red

Beefwood, *Grevillea striata*

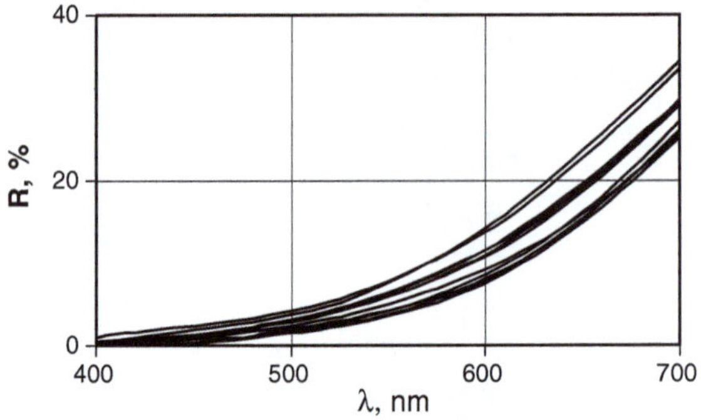

$L_0^* = 36 - 25$ %, $R_m = 12 - 7$ %, $\lambda_0 = 582 - 593$ nm, $h^* = 2.2 - 1.6$, $K = 0.37 - 0.24$, Colour hue: br.red-red

Swamp Kauri, *Agathis australis*

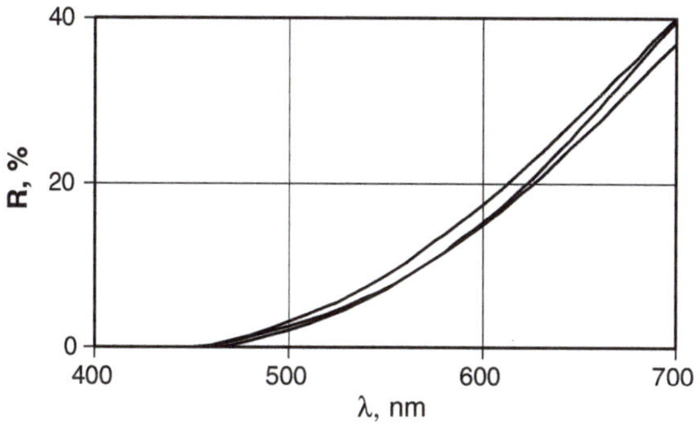

$L_0^* = 37 - 34$ %, $R_m = 14 - 13$ %, $\lambda_0 = 570 - 573$ nm, $h^* = 2.6 - 2.4$, $K = 0.53 - 0.49$, Colour hue: brown

Sassafras, *Atherosperma moschatum*

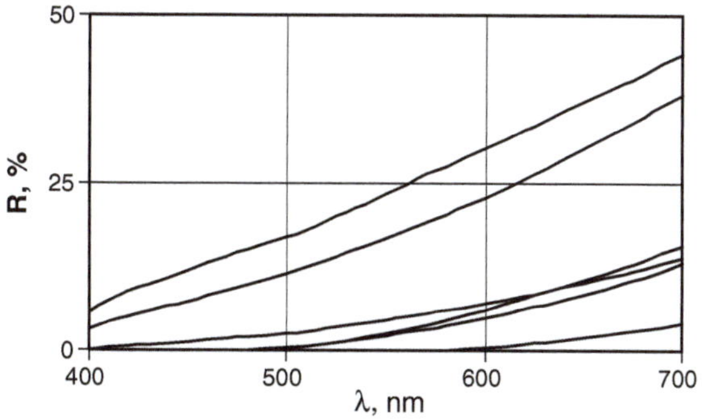

$L_0^* = 55 - 2$ %, $R_m = 24 - 0.8$ %, $\lambda_0 = 555 - 612$ nm, $h^* = 3 - 0.31$, $K = 0.83 - 0.1$, Colour hue: yellow-red

Index

E. Csanády et al., *Quality of Machined Wood Surfaces*,
DOI 10.1007/978-3-319-22419-0

MIX
Papier aus verantwortungsvollen Quellen
Paper from responsible sources
FSC® C105338

If you have any concerns about our products,
you can contact us on
ProductSafety@springernature.com

In case Publisher is established outside the EU,
the EU authorized representative is:
Springer Nature Customer Service Center GmbH
Europaplatz 3, 69115 Heidelberg, Germany

Printed by Libri Plureos GmbH
in Hamburg, Germany